ORNAMENTAL CROPS AND LANDSCAPING
PRODUCTION TECHNOLOGY

ORNAMENTAL CROPS AND LANDSCAPING
PRODUCTION TECHNOLOGY

V RAVI
RANDEEP N C

KNOWLEDGE
BAKERS

KNOWLEDGE BAKERS

ISBN: 978-93-49964-99-0

Published by : **Knowledge Bakers**
Pune, India
info@knowledgebakers.com

PREFACE

Ornamental Crops and Landscaping: Production Technology offers a comprehensive examination of the scientific principles, production practices, and aesthetic applications associated with ornamental horticulture and landscape design. In response to increasing urbanization, environmental awareness, and demand for sustainable green spaces, the cultivation of ornamental plants and the integration of landscaping have taken on new relevance in both rural and urban contexts.

This book is intended for students, researchers, horticulturists, landscape architects, and professionals in the floriculture and landscaping industries. It aims to bridge the gap between theoretical knowledge and practical application by presenting contemporary production technologies, innovative design approaches, and evolving trends in the ornamental plant sector.

The early chapters provide an overview of ornamental plant classification, propagation techniques, and nursery management. Readers are introduced to foundational concepts in plant physiology, soil and nutrient management, and integrated pest and disease control, tailored specifically to ornamental species. Special attention is given to climate-specific cultivation practices, varietal selection, and the role of plant breeding in developing aesthetically superior and stress-tolerant cultivars.

Subsequent sections focus on landscape design principles and the functional use of ornamental plants in residential, institutional, recreational, and urban settings. The book covers essential design elements—such as symmetry, texture, color, and scale—alongside practical considerations for hardscape integration, irrigation systems, and sustainable landscaping. Case studies illustrate how aesthetic value, ecological balance, and environmental services can be harmoniously achieved through strategic planting and design.

Advancements in protected cultivation, vertical gardening, rooftop landscaping, and the use of automation in ornamental crop management are also discussed, reflecting the shift toward high-efficiency, space-optimized systems. The role of floriculture in the global economy, including the cultivation and post-harvest management of cut flowers, foliage, and potted plants, is explored with a focus on quality assurance and market demands.

Furthermore, the book examines the environmental benefits of ornamental horticulture, including air purification, biodiversity enhancement, microclimate regulation, and mental well-being. In doing so, it underscores the importance of green infrastructure and the role of professionals in promoting sustainable urban development.

Through its integrated approach, this volume seeks to equip readers with both the scientific knowledge and the practical tools required to thrive in the dynamic field of ornamental crop production and landscape design. By highlighting recent innovations and real-world applications, it fosters a deeper understanding of how horticultural beauty and ecological function can coexist and contribute to improved quality of life.

V Ravi

Randeep N C

CONTENTS

Chapter 1

IMPORTANCE AND SCOPE OF ORNAMENTAL CROPS

1.1 INTRODUCTION

Ornamentals, in horticulture, include both woody and herbaceous plants used primarily as amenities. Ornamentals, in horticulture, include both woody and herbaceous plants used primarily as amenities. Woody ornamentals are generally perennial (ie, persist for more than 2 years) and include trees, shrubs and vines; herbaceous ornamentals are annual, biennial or perennial (persist for one, 2, or more years) and make up a vast array of plants that provide colour in flower beds and borders. In winter hardiness is very important for growing ornamentals. The ability of a woody plant to survive a winter depends largely on the species geographic origins, the stage of maturity reached, and the resulting cold resistance achieved by the end of the growing season. Plants that require a long season to mature seldom achieve this state when grown in regions with short growing seasons.

Winter hardiness does not present the same problem with many herbaceous ornamentals, as many of them are annuals and do not overwinter. When the normal growing season is not long enough for them to grow and produce flowers, the grower often starts them indoors (eg, in a greenhouse), and transplants them outside when the risk of frost is over. The common snapdragon, for example, is generally seeded in the greenhouse in February and planted outside in late May or early June. If such plants were seeded directly outdoors in May, in regions where the frost-free period is 90 days or less, they would blossom for only a few weeks before being killed by frost. Herbaceous biennials and perennials are often allowed to remain in the soil all year; provided that they are sufficiently cold tolerant. Their winter survival is generally assured if they receive adequate snow cover to protect them from severe temperatures and prevent alternate freezing and thawing of the soil around the roots. Some herbaceous biennials and perennials, such as gladiolus, are not winter hardy and must be dug up and stored if they are to be grown as a perennial in our climate. Otherwise, they can be grown as annuals.

Generally, ornamentals are either selected forms of specific plants (ie, cultivars a.k.a. cultivated varieties) or hybrids between species. Because of the winter-hardiness problem, many woody ornamental cultivars and hybrids have been derived from plants indigenous to Canada and the northern US, or from regions with climates similar our own such as China, Siberia, and parts of Europe. Herbaceous ornamentals are also often made up of cultivars of species and hybrid cultivars. Improvement programs involving hybridization, both within and between species, are commonplace. Hardy perennials and biennials are chiefly temperate-region plants while annuals can be derived from many different climates and geographic regions.

Ornamental Crops are grown around the school compound or home surroundings to make them look beautiful. Examples:

1. Lawn grass
2. Lantana Flower
3. Hibiscus Flower
4. Ice Plant
5. Sunflower
6. Flame of the forest flower.
7. Croton
8. Lily Flower
9. Rose Flower
10. African Marigold
11. Euzora
12. Cockscomb

1.2 IMPORTANCE

Ornamental plants are mostly grown for their exquisite blooms and are a source of major attraction for many gardens. Several such ornamental gardens usually prefer a wide variety of flowering plants so that the garden is continuously in flower through the year during spring, summer, monsoon and winter. Several types of plants representing predominantly angiospermic plant families, some selected gymnosperms and pteridophytes (such as ferns) are most commonly grown that have colorful flowers, foliages, shapes, fragrance or aroma, spectacular morphological characters that are visibly attractive are usually selected. Ornamental plants could be either herbs, shrubs or tress, lianas and creepers, terrestrial and aquatic in habit and that could have originated in a wide variety of habitats representing different ecosystems and environments. Several ornamental plants are commonly known or their exquisite beauties such as gardenia, pansy, bougainvilleas, mussaendas, roses, narcissus, daisies, petunias, dahlia, chrysanthemum, gerbera, aster, cosmos, gladiolus, sunflowers, pansy, different lilies, amarylids, orchids colorful ornamental grasses and aroids. In general, floriculture and nursery industries are often referred together as ornamentals or ornamental industry or non-food agriculture. Recently the sod and Christmas tree industries have been referred to as ornamental horticulture in the Europe and North America; while use of plants in reshaping and changing landscape profile, quality and property value is now called ornamental landscaping.

In addition to herbs, shrubs and trees, miniature plants or bonsai, cacti and succulents also constitute ornamental plants. It is important to understand that ornamental plants are not just restricted to only spectacular flowering plants. The concept of ornamental plant is also a dynamic. Hence, a weed in one part of a continent could be ornamental in another part of the same or a distant continent. A commercial food or industrial crop with spectacular foliage or bloom in a specific country or continent may be considered as an ornamental species in another distant locality. Many producers prefer ornamental plants with a special emphasis for food production. However, it is possible to integrate floral and fruit productions together in the growing regions by targeted selection of specific plant species. For example plants like Runner Beans, Ruby Chard or Globe Artichokes cab easily find their place inside the walls of a floral garden for ornamental plants. People can decide about the merits of including such plants in their ornamental gardens as they can add a new dimension to their floral gardens, lawns and parks.

Roadside ornamentals along the busy city streets and lanes and along the highways passing through both urban municipalities or corporations and rural areas are an important source of beauty and aesthetics for travelers and visitors. Well maintained city and town parks are also added attraction for tourist to these places and also have a positive impact on the local environment. It is also important to note that several such parks, gardens and ornamental meadows serve as island refuge for many small mammals and numerous bird species that adds to the color of life and helps in protecting the local biodiversity particularly in busy urban commercial or business centers as well as residential areas. Ornamental plants provide homeowners the means to improve their local environment. It is therefore important to have the necessary knowledge and technology of selecting, multiplying and growing appropriate species of plants that would be most suitable to fit into the target landscape. Overall, ornamental plants serve multiple roles to the human society. In one hand they enrich the aesthetic quality of our life and improve the local environment providing refuge for our busy life and relaxation; while on the other, they protect the nature, local biodiversity, generates revenues and employment and helps in the process of natural beautification. In addition to wide variety of colorful plants and plants showing distinct foliages for pattern, color, shape, size and morphology; artificial miniaturized plants or "bonsai" also constitute an important aspect of ornamental plants are exquisite parts of indoor beautification with plants.

Although several ornamental species are known, only a few distinct handfuls are described as example below.

Bougainvilleas

Bougainvilleas represent woody climbing plants belonging to the dicotyledonous family Nyctaginaceae and are adored for their wide spectrum and shades of colors across the globe. It is however important to note that the different shades of color for which different cultivars and germplasms of Bougainvilleas are so revered and cherished all over the world are not actually true petals. But truly speaking these are colorful bracts that translates in simple terms to mean modified leaves. The true flowers of the plants are small, tubular, white and subtended by the colorful bracts.

The flowers occur in clusters of three with 3-6 colorful bracts surrounding them. The spines of on the stem plant are solid, green when immature but transforms to become woody at maturity. Leaves are simple, ovate and alternately arranged on the stem with a thick leathery texture. Now horticulturists have been successful in introducing many varieties of the plants with variegated leaves that has further enhanced the ornamental value of the plants in addition to the colorful bracts. Different species have different colored bracts varying between purple, pink, red, orange to white and yellow and several other shades. The plant is exploited commercially as an important ornamental and horticultural plant and several hybrids and mutant varieties are also well known.

The center of origin of this spectacular garden plant is tropical South America. The strange name of the plant comes from the last name of the celebrated explorer and polymath genius Captain Louis Bougainvillea. The plant was actually named by the famous French botanist Philbert Commerson. It is worth mentioning that bougainvillea is the national plant of Grenada, a small island nation from the Caribbeans. The plant is easy to cultivate and do not need much maintenance once a steady growth stage is reached. The bright colored bracts against the green leaves usually demonstrate an amazing color contrast that adds to the natural beauty of the bougainvillea hedge or ornamental arrangements of the growing and well pruned plant around garden gates and against boundary walls or against any supporting structures around which the plant can creep around and grow successfully. The plant has been found to be quite resistant against city based pollutants and have seen to thrive under bright sunny conditions with low water requirements. Hence this is an excellent city ornamental that could cope with pollutants and continue to add to the city aesthetics with the spectacular color of their bracts.

Cacti and Succulents

There are several plant species that adorn our homes and add to the indoor beauty in the environment that we settle in; our homes and offices. Even in absence of hot houses or greenhouses or so called glass houses such plants could be reared and easily grown and cultivated in the ambient temperatures of our houses and offices. Such plants include several xeric and succulents plants from angiospermic or true flowering groups of plants such as cactus (plural cacti), several members of bromeliads, euphorbias, lilies, amaryllids, asperagales, aroids, certain hardy artificial hybrid orchid species, short indoor palms and bamboos, bonsai plants (specially prepared live plant miniatures) etc; and non-flowering plants such as ferns adorn the inner world of modern human habitation across the planets mostly in towns and cities. However, cacti and succulents constitute a big bulk of them and are represented by a bunch of diverse plant families.

Over 15-16,000 species of cacti are known worldwide and include big desert giants to minute potted ones showing wide diversity in their habits and habitats; while the succulents plants have highly specialized and adopted thick and fleshy leaves and stems that are capable of retaining moisture inside the plant cells in xeric (arid/desert) habitats. Both cacti and other succulents can store and preserve moisture efficiently for long periods of time and unfavorable conditions of temperature and humidity. Hence such plants are prized as indoor-ornamental plants. Such plants have evolved several specialized modified plant parts such as flattened, thick fleshy leaves and stems, specialized cuticle and waxy coatings, spines and thorns, sunken stomata and several other morphological and anatomical macro and micro structures to enable them to survive them under hostile environment. As such due to low water requirements and their ability to survive under minimal conditions such plants have become so popular worldwide as fancy indoor plants beautifying our surroundings and also helping in absorbing several indoor pollutants and purifying the enclosed environments of houses and offices. Such diverse pants not only add aesthetic beauty but also have a close ecological relationship to our anthropocentric life and society. The wide adaptability of such plant groups to their hostile environment and ecosystem has been through long years of evolution and are spectacular in the very true sense of the term.

Orchids:

Orchids are monocotyledonous plants belonging to the family Orchidaceae. The largest member of the plant family is found in the tropics and sub-tropics of Asia, Oceania, Africa and the Americas and also in the temperate regions of Europe and

Asia. The family is represented by diverse plants with a wide spectrum of shape, forms and color and as such highly valued as horticultural plants. The orchids therefore constitute an important member of plants categorized as highly prized ornamentals. Orchids (like tulips), have enormous potential as ornamental plants in the local, regional and international markets and serve as important source of foreign exchanges for many developing and under developed countries.

Orchid

Flowers have been considered as the symbol of grace and elegance and a feast for our eyes. They are used on all religious festival occasions. Flowers are given as a birthday presents, wedding gifts or while meeting sick people and even at funerals. Most Hindu ladies adhere their; hair style with flowers i.e. Gajara and Veni and it is one of the important floral ornament which will and grace to their beauty. All the people irrespective of their origin, race, sex and cadre love flowers.

Generally flowers are offered by devotees in Temple, Gurudwara, Church and Masjids- Flowers are used as a floral decoration. Even dried flowers are also used in flower craft or arrangement of garlands and bouquets are prepared and offered to welcome the dignitaries. When cut flowers are used for vase decoration, that will be a marvelous piece of indoor decoration.

Importance of flowers is not restricted upio the beautification, decoration or preparation of Gajra, Garland, Veni or Bouquets but afso have the industrial importance too. Some flowers like Rose, Jasmines, Tuberose, Kevda, Bakul are used for extraction of essential oils which is base for preparation of perfumes, scents or attar. From rose Gulkand, Rose water etc. products are also prepared.

1.3 SCOPE

There is a good scope for commercial floriculture. The important factors which decide the scope for Commercial Floriculture are Soil, Climate, labour. Transport and Market. All most all big cities are developing very speedily to accommodate this fast growing population, cement concrete, jungle is also developing at some rate and thus peoples are now realizing the importance of open space, parks and garden for relaxation, peace of mind, recreation and unpolluted air. Thus, to meet out all these problems bio-aesthetic planning is essential, which runs hand in hand with town planning. In modem life floriculture garden in the country yard is an integral part of the modern life and thus ornamental plants has find a pride of place in home gardening.

After liberalization, the Govt. of India identified floriculture as a sunrise industry and accorded it 100 percent export oriented status. Owing to steady increase in demand of flower floriculture has become one of the important Commercial trades in Agriculture. Hence commercial floriculture has emerged as hi-tech activity-taking place under controlled climatic conditions inside greenhouse. Floriculture products mainly consist of cut flowers, pot plants, cut foilage, seeds bulbs, tubers, rooted cuttings and dried flowers or leaves. The important floricultural crops in the international cut flower trade are rose, carnation, chrysanthemum, gargera, gladious, gypsophila, liastris, nerine, orchids, archilea, anthuriu, tulip, lilies.

Present status and growing trade is still in infancy. Floriculture in India, is being viewed as a high growth Industry. Commercial floriculture is becoming important from the export angle. The liberalization of industrial and trade policies paved the way for development of export oriented production of cut flowers. The new seed policy had already made it feasible to import planting material of international varieties.

The government of India offers tax benefits to new export oriented floriculture companies in the form of income-tax holidays and exemption from certain import duties. Agricultural and Processed Food Products Export Development Authority (APEDA), responsible for export promotion and development of floriculture in India, grants subsidies for establishing cold storage, precooling units, refrigerated vans and green houses, and air freight subsidy to exports. It has been found that commercial floriculture has higher potential per unit area than most of the field crops and is therefore a lucrative business.

According to a report of the APEDA, the total area under flower crops was estimated around 34,000 hectares, which included 24,000 hectares under traditional flowers such as marigold, jasmine, aster, rose, chrysanthemum, tuberose and 10,000 hectares under modern flowers like cornation, rose, gerbera, gladiolous, anthurium.

Returns from floricultural products were estimated at Rs.205 Crores, which included Rs.105 Crores from traditional and Rs. 100 Crores from modern flowers. The data collected by National Horticultural Board show that estimated area and production were 53,000 hectares and 4 lakh tonnes. In 1995-96, there were mainly 230 export-oriented units in India. More than 50% of the floriculture units are based in South zone mainly in Karnataka, Andhra Pradesh, Tamil Nadu. Also West Bengal, Maharashtra, Rajasthan have large areas under floriculture. The domestic flower production goes on increasing annually. Technical collaborations with foreign companies have been approved for India, in order to increase total share in the floriculture world trade.

As far as flower trade is concerned i.e. for cut flowers and loose flowers, it is growing very well in our state because these cut flowers are used for vase decoration and now-a-days there is a craze for indoor decoration. As far as loose flowers are concerned these are mainly used for preparation of gajara, veni, garland and bouquets and thus demand of flowers for these purpose is unending. Thus, taking into consideration the different points i.e. bio-aesthetic planning, floral garden, indoor decoration, social functions and religious functions the demand for floricultural plants is increasing day by day and to meet out the same there is a good scope for growing and raising of Ornamental or Floricultural plants. When Flower Trade is concerned; different flowers like Rose, Chrysanthemum, Gladiolus, Tuberose are demanded in the market as cut flowers. While Aster, Gaillardia, Marigold, Chrysanthemum, Jasmines, Tager Nerium as loose flowers.

Almost all cities are developing very speedily to accommodate fast growing population. They build more cemented concrete structure and polluting our living environment. Thus, peoples are now realizing the importance of open space, parks and garden for relaxation, peace of mind, recreation and fresh air. To solve these problems bio-aesthetic planning is essential, which runs hand in hand with town planning.

1.3.1 Favourable Factors

Climate

The climate and soil should be suitable for growing flower and ornamental plants. Ornamental plant production in greenhouses, like most businesses, is planned according to a unit sales forecast. The number of plants is easily planned, but that is where the challenge of working in a biological industry begins. Plants, simply put, are biochemical reactors that are subject to chemical reaction rates. The popular rule of thumb for biological systems is that reaction rates double for every 10°C of temperature rise, up to a limit. So, temperature dictates plant mass, or quantity. However, quantity of ornamental plants does not guarantee profitable sales without sufficient quality. Quality is a short-lived aspect of plant production that is critical to profits, and the greenhouse climate is critical in determining crop quality.

Quantity and quality, alone or together, do not guarantee profit without timing. Hitting market windows is critical. We all know that the spring weekends are precious and that there are no New Year's markets for the poinsettia. Heating, cooling, ventilation and air distribution combine to result in a climate that is confined to a rather narrow range of conditions. This environment, hospitable for plants yet hostile to pathogens, is crucial for success. Greenhouse climate screens, or energy/shade curtains as they are frequently known, are frequently installed for insulation to reduce heating and cooling costs. However, recent research reinforces the benefit of curtains in creating a climate more favorable to plant production.

The rate of plant development is dictated by the mean daily temperature of the shoot tip. In 2001, research at Michigan State University demonstrated that campanula time to flower was reduced by about two weeks by increasing the temperature of the shoot tip from 16 to 24°C. Recently it was reported that growing New Guinea impatiens underneath a greenhouse curtain raised the shoot tip temperature by 0.5 to 2.3°C as a result of increasing the long wave radiation to the plant, which was 70 to 125 percent greater than the case with no curtain in a glasshouse.

Cheap Labour

Cheap and available labour is a positive side of any country for cultivating flower and ornamental plants which definitely helps in keeping down the production cost. Growers need to evaluate their production costs and plant inventory to determine if adjustments need to be made to stay competitive. To accomplish this, growers need to have a good handle on their costs. MSU Extension has developed inexpensive software to help you evaluate your nursery business from a financial perspective. "Estimating the Wholesale Cost of Nursery Production" is an Excel spreadsheet type of software tool that allows you to use your records to estimate your production costs and to explore opportunities for cost savings.

Demand

Flowers are using for vase decoration, indoor decoration, preparation of gajara, veni, garland and bouquets and thus demand of flowers for these purposes is unending. Ornamental plant retailers face multiple challenges, including identifying their market, analyzing their current and potential consumer base, determining prices, and developing promotional programs to attract new customers. Previous research has focused on investigating consumer demographic profiles as well as consumers' store choices and their preferences for ornamental plant attributes, such as colors and scents. Research investigating consumption patterns such as seasonal and occasional purchases, repeat-buying habits, and purchase frequency has also benefited the industry by providing useful insights for developing effective marketing strategies. Research in other industries shows that consumers react to incentives, such as relevance and benefits of products and services provided. For example, the food industry is active in promoting the nutritional benefits of their products through various marketing channels, such as labels and advertisement. Research studies found that the effectiveness of these strategies was significant.

There is a great increase in demand of floricultural products with increasing income and globalization of economy. Netherlands, Italy, Germany and Japan have strong tradition for growing and consumption of flowers.

However new production centers are developing in Latin America-Columbia, Costa Rica, Chile; Africa-Kenya, Rhodesia Morocco, South Africa etc. Asia-Israel, India, Sri Lanka, etc. to meet the internal demand as well as export. These countries produce quality product at low cost, off course, with high cost of investment.

Export

Considering the bio-aesthetic planning, floral garden, indoor decoration, social functions and religious functions the demand for flowers are increasing day by day in the world market and there is a good scope to export. If we can produce quality flowers then there is a wide scope for exporting. The ornamental sector includes the floriculture (primarily cut flowers and potted plants), nursery, Christmas tree, and turf sod industries. In 2016 the sector generated farm gate sales of $2.3 billion, including resales by producers. Floriculture remained the largest revenue generating sub-sector, accounting for 65.9% of total sales. Floriculture sales increased to $1.51 billion in 2016, up 0.9% from 2015 and 10.9% from 2012.

Ornamental products are distributed through multiple channels. Mass-market chain stores are the primary distributers of floriculture products, accounting for 31.5% of total sales in 2016. However, landscape contractors and garden centres are the two largest distribution channels specifically for nursery products, making up 26.3% and 19.6% of total sales, respectively.

In spite of an abundant and varied production base, India's export of floricultural product is not encouraging. The low performance is attributed to many constraints like non-availability of air space in major airlines, since most of the airline operators prefer heavy consignments. The existing number of flights during the peak seasons is not sufficient for export purpose.

Exporters for infra-structural problems like bad interior road, inadequate refrigerated transport and storage facilities. Lack of professional backup of delivery and supporting companies, which resort into high cost of technology for Indian entrepreneurs. Tedious Phyto-sanitary certification and an unorganized domestic market. In order to overcome these problems, attention must be focused on:

- ☆ Reduction in import duty on planting material and equipment.
- ☆ Airfreight should be reduced to a reasonable level. Sufficient cargo space may be provided in airlines.
- ☆ Establishment of model nurseries far supplying genuine planting material.
- ☆ Co-operative florist organizations should be established at regional level.
- ☆ Training centres for diploma course on the pattern of ITI for training the personnel in floriculture should be set up.
- ☆ Exporters should plan and monitor effective quality control measures right from production to post harvesting, storage, and transportation.
- ☆ An analysis of strengths, weaknesses, opportunities of the floricultural industry shows that India has immense potential for export of floricultural products.
- ☆ Though the global floriculture industry is growing comparatively at a faster pace than in India, still a scope exists to bridge the demand and supply gap.

1.4 USES

1.4.1 Introduction

Ornamental plants are used to provide greenery in cities and other inhabited areas, in gardens and parks, and outside of public buildings and residences. They are distinct because of their attractive shapes and the different colors of their flowers, leaves, and fruits. In terms of biological properties and agro technical requirements, ornamental plants are divided into several groups, such as trees and shrubs, perennials, biennials, annuals, grasses, and bulbs.

Ornamental trees and shrubs include leafy and conifer types, evergreens, and deciduous plants. They are used in parks and squares and to line boulevards and streets. In separate plantings, varieties with spreading crowns (oak, ash, or plane trees) or umbellate crowns (silk trees) are used. Plants with weeping crowns (weeping willow, weeping birch) serve to decorate bodies of water. In alpine terrain, on slopes, and on steep mountains, spreading plants are especially attractive (rock cotoneaster, mountain pine, juniper). To bring greenery to streets and form treelined boulevards, trees with pyramidal crowns (cypress, poplar), round crowns (false acacia, linden), or conical crowns (fir, spruce) are planted. Many treelike plants can be shaped in any way through pruning; these include the box, thuja, laurel, yew, hornbeam, hawthorn, elm, and privet. Creeping plants (ivy, clematis, grape, ampelopsis) serve as vertical vegetation on walls, terraces, and belvederes; they also decorate architectural constructions such as arbors, trellises, and columns. Besides the local species, exotic ones imported from other regions are quite common.

Ornamental perennials are herbaceous plants that live for more than two vegetative periods. This is the richest group of ornamental plants in terms of species, and it includes flowering plants, plants with ornamental leaves, and plants with ornamental fruits. Ornamental perennials are used widely in gardens, parks, and squares. In temperate climates perennials are subdivided into plants that winter outdoors and those that must be taken inside for the winter. The latter are kept as starting stock, tubers, and rhizomes in dry storage at a temperature of 5°C. Certain perennial plants of southern origins are used as annuals in more rigorous climates (including ageratum, petunia, scarlet sage, and nasturtium). Perennials are regrown in the spring from rhizomes, tubers, or bulbs through restorative buds. According to their ornamental features, they are classified into different types. Flowering types are distinguished by abundant blossoms, brilliant colors and shapes, doubleness of flowers, and the aroma of the blossom or mass of small blossoms gathered into a raceme. Ornamental foliage plants are classified according to leaf slits, color variegation, brilliance of hues, and evergreens. Other classifications are plants with beautiful, brightly colored fruits; plants that develop exquisite brooms or heads (the grains); compact plants that form a turfy cushion as they grow, and the lianas (creepers, climbers, or crawlers). Perennials are used to create ornamental arrangements in flower beds, partitions in pavements, lawns, meadows in parks, and sunny clearings. Sometimes perennials are planted as solitaries, with only a single variety of large, highly ornamental plant. Perennials are used as colorful ground covers in flower beds and borders; creepers are used for intricate structures such as belvederes, arbors, arches, and curtains of greenery. Many perennial blossoms can be used in bouquets. They grow in one spot for three to five years or longer and multiply by division, shoots, or seed plantings.

Biennials are flowering ornamental plants which in the first year of growth develop foliage and in the second year produce stems with flowers and fruits. After maturing, they die. This group sometimes includes some perennials that are cultivated as biennials. True biennials include the alpine forget-me-not, which is used in flower beds on open soil and also for cut flowers. The Canterbury bell is used in flower beds and in open soil; honesty, in dry bouquets; foxglove, in group plantings and for cut flowers; and the mallow or hollyhock, in group plantings and for cut flowers. Of the herbaceous perennials that are cultivated chiefly as biennials, the best known are pansies and sweet William, which are grown on open soil; hesperis, or dame's violet, used for group plantings, cut flowers, and distillation; and daisies, used for planting in borders and flower beds and, in winter, for growing indoors in pots. Biennials multiply by seed plantings in hothouses or on open soil, and some send out shoots.

The general term "ornamental annuals" is used for ornamental plants that are raised from seeds in the course of a single growing season. Most annuals are plants that grow for only a single year, but biennials and perennials that blossom in one year but do not winter outside (heliotrope, snapdragon, fuchsia, and lobelia) are considered as ornamental annuals too. Annual plants produce groups of lovely blossoms, ornamental leaves (amaranth, ornamental cabbages, and castor-oil plants), dried flowers (everlastings), cover plants, vines, hanging plants, and bouquet flowers. In temperate climates, most annuals are planted in March or April in hothouses and greenhouses in order to lengthen their growing season or to encourage earlier flowering. Later they are transplanted outdoors. Many annuals are planted directly in soil outdoors. The annuals include certain asters, larkspurs, balsams, pot marigolds, stocks, poppies, nasturtiums, petunias, tobaccos, marigolds, morning glories, sweet peas, and mignonettes. They are used for flower beds, mixed borders, borders, boxes, hanging pots, and group plantings.

Ornamental grains are grain plants used in ornamental gardening for borders and group plantings and for making bouquets of live flowers as well as dried ones. For individual and group plantings by walks, high grains are used (reedy arundo, red eriantus, wild sugarcane, and pampas grass); for borders and group plantings, short grains are used (canary grass, fescue, hare's-tail grass, and tall ryegrass); for walks, meadow grass, fescues, ryegrass, and bent are used.

Bulb ornamental plants are of the Liliaceae and Amaryllidaceae families and are cultivated in ornamental gardening: snowdrops, squills, tulips, narcissus, hyacinths, and lilies. Hyacinths, tulips, narcissus, and some other bulb types successfully multiply by forced planting, that is, in hothouses and rooms. Bulb culture is developed in southern parts of the USSR, such as the humid subtropical zone of the Black Sea shore and the southern shore of the Crimea. In open plantings in the central regions of the RSFSR, bulbs such as snowdrops, squills, and narcissus are used chiefly for decoration in perennial flower beds and borders. In the southern part of the USSR, bulbs winter outdoors without cover, whereas in the central regions some plants, such as hyacinths, are covered with peat or fallen leaves in the winter. Bulbs multiply mainly by tubers that appear on the old bulb or by squamules or outer divisions of the bulb. Often tuberous plants are considered as bulbous ornamental plants, since they are similarly cultivated and their underground organs are outwardly similar. They include gladiolus, saffron, and meadow saffron.

1.4.2 Features

1. Ornamental trees provide shade for pedestrians on hot days, decreasing the use of vehicles. On colder days, the trees bulk up and can reduce frost by retaining heat in the soil.
2. They absorb carbon dioxide from the air and output oxygen; creating cleaner air. This is especially important in cities that are prone to pollution. When

used indoors, they help purify the air and create a pleasant atmosphere. The peace of lily may be an example

3. Outdoor ornamentals help beautify the landscape and create an appealing environment. Fallen leaves make good mulch and compost.
4. They are grown for food. Some chilies are grown as ornamental plants. An example is capsicums or Doret Nag, officially the hottest chili in the world.
5. Ornamental Plants and trees planted at the sides of the roads help drivers reduce stress. It also protects drivers from severe car accidents by serving as a barrier to prevent cars from somersaulting or other worse cases.
6. If a person knows how to grow ornamental plants, he or she could earn some money by selling grown ornamentals. Since ornamentals sold at green houses or garden stores can be expensive.
7. Some ornamentals can be used for medicinal purposes such as rosemary, mexican marigold and lemon grass.
8. Ornamental plants can brighten a yard, compliment the style of a house or building. Therefore increasing its worth by a percentage; making more money for the owner or portraying the owner of a high standards.
9. The foliage of evergreen Ornamental trees brightens winter landscape and affords windbreaks that save money on fuel bills.
10. Ornamental gardens provide a relaxing, protective environment which offers privacy to its users.
11. Ornamentals can be used for decorations during occasions such as weddings and anniversaries. A common example is rose.

Chapter 2

MEDICINAL AND AROMATIC PLANTS

2.1 EXTRACTION OF MEDICINAL PLANTS

Extraction, as the term is used pharmaceutically, involves the separation of medicinally active portions of plant or animal tissues from the inactive or inert components by using selective solvents in standard extraction procedures. The products so obtained from plants are relatively impure liquids, semisolids or powders intended only for oral or external use. These include classes of preparations known as decoctions, infusions, fluid extracts, tinctures, pilular (semisolid) extracts and powdered extracts. Such preparations popularly have been called galenicals, named after Galen, the second century Greek physician.

The purposes of standardized extraction procedures for crude drugs are to attain the therapeutically desired portion and to eliminate the inert material by treatment with a selective solvent known as menstruum. The extract thus obtained may be ready for use as a medicinal agent in the form of tinctures and fl uid extracts, it may be further processed to be incorporated in any dosage form such as tablets or capsules, or it may be fractionated to isolate individual chemical entities such as ajmalicine, hyoscine and vincristine, which are modem drugs. Thus, standardization of extraction procedures contributes significantly to the final quality of the herbal drug.

2.1.1 Processes of Extraction

Maceration

In this process, the whole or coarsely powdered crude drug is placed in a stoppered container with the solvent and allowed to stand at room temperature for a period of at least 3 days with frequent agitation until the soluble matter has dissolved. The mixture then is strained, the marc (the damp solid material) is pressed, and the combined liquids are clarified by filtration or decantation after standing.

Infusion

Fresh infusions are prepared by macerating the crude drug for a short period of time with cold or boiling water. These are dilute solutions of the readily soluble constituents of crude drugs.

Digestion

This is a form of maceration in which gentle heat is used during the process of extraction. It is used when moderately elevated temperature is not objectionable. The solvent efficiency of the menstruum is thereby increased.

Decoction

In this process, the crude drug is boiled in a specified volume of water for a defined time; it is then cooled and strained or filtered. This procedure is suitable for extracting water-soluble, heat-stable constituents. This process is typically used in preparation of Ayurvedic extracts called "quath" or "kawath". The starting ratio of crude drug to water is fixed, e.g. 1:4 or 1:16; the volume is then brought down to one-fourth its original volume by boiling during the extraction procedure. Then, the concentrated extract is filtered and used as such or processed further.

Percolation

This is the procedure used most frequently to extract active ingredients in the preparation of tinctures and fluid extracts. A percolator (a narrow, cone-shaped vessel open at both ends) is generally used. The solid ingredients are moistened with an appropriate amount of the specified menstruum and allowed to stand for approximately 4 h in a wellclosed container, after which the mass is packed and the top of the percolator is closed. Additional menstruum is added to form a shallow layer above the mass, and the mixture is allowed to macerate in the closed percolator for 24 h. The outlet of the percolator then is opened and the liquid contained therein is allowed to drip slowly. Additional menstruum is added as required, until the percolate measures about three-quarters of the required volume of the finished product. The marc is then pressed and the expressed liquid is added to the percolate.

Sufficient menstruum is added to produce the required volume, and the mixed liquid is clarified by filtration or by standing followed by decanting.

Hot Continuous Extraction (Soxhlet)

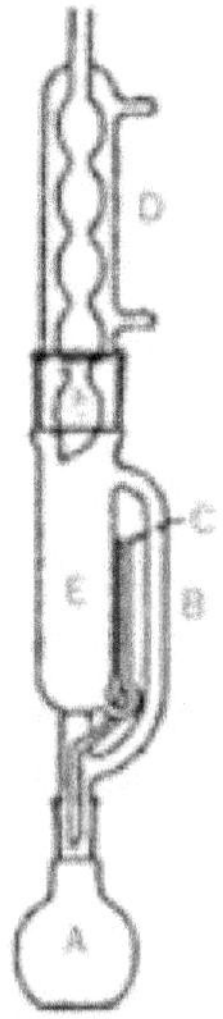

Soxhlet apparatus

In this method, the finely ground crude drug is placed in a porous bag or "thimble" made of strong filter paper, which is placed in chamber E of the Soxhlet apparatus (Figure). The extracting solvent in flask A is heated, and its vapors condense in condenser D. The condensed extractant drips into the thimble containing the crude drug, and extracts it by contact. When the level of liquid in chamber E rises to the top of siphon tube C, the liquid contents of chamber E siphon into fl ask A. This process is continuous and is carried out until a drop of solvent from the siphon tube does not leave residue when evaporated. The advantage of this method, compared to previously described methods, is that large amounts of drug can be extracted with a much smaller quantity of solvent. This effects tremendous economy in terms of time, energy and consequently financial inputs. At small scale, it is employed as a batch process only, but it becomes much more economical and viable when converted into a continuous extraction procedure on medium or large scale.

Aqueous Alcoholic Extraction by Fermentation

Some medicinal preparations of Ayurveda (like asava and arista) adopt the technique of fermentation for extracting the active principles. The extraction procedure involves soaking the crude drug, in the form of either a powder or a decoction (kasaya), for a specified period of time, during which it undergoes fermentation and generates alcohol in situ; this facilitates the extraction of the active constituents contained in the plant material. The alcohol thus generated also serves as a preservative. If the fermentation is to be carried out in an earthen vessel, it should not be new: water should first be boiled in the vessel. In large-scale manufacture, wooden vats, porcelain jars or metal vessels are used in place of earthen vessels. Some examples of such preparations are karpurasava, kanakasava, dasmularista. In Ayurveda, this method is not yet standardized but, with the extraordinarily high degree of advancement in fermentation technology, it should not be diffi cult to standardize this technique of extraction for the production of herbal drug extracts.

Counter-current Extraction

In counter-current extraction (CCE), wet raw material is pulverized using toothed disc disintegrators to produce a fine slurry. In this process, the material to be extracted is moved in one direction (generally in the form of a fine slurry) within a cylindrical extractor where it comes in contact with extraction solvent. The further the starting material moves, the more concentrated the extract becomes. Complete extraction is thus possible when the quantities of solvent and material and their fl ow rates are optimized. The process is highly efficient, requiring little time and posing no risk from high temperature. Finally, sufficiently concentrated extract comes out at one end of the extractor while the marc (practically free of visible solvent) falls out from the other end.

This extraction process has signifi cant advantages:

1. A unit quantity of the plant material can be extracted with much smaller volume of solvent as compared to other methods like maceration, decoction, percolation.
2. CCE is commonly done at room temperature, which spares the thermolabile constituents from exposure to heat which is employed in most other techniques.
3. As the pulverization of the drug is done under wet conditions, the heat generated during comminution is neutralized by water. This again spares the thermolabile constituents from exposure to heat.
4. The extraction procedure has been rated to be more efficient and effective than continuous hot extraction.

Ultrasound Extraction (Sonication)

The procedure involves the use of ultrasound with frequencies ranging from 20 kHz to 2000 kHz; this increases the permeability of cell walls and produces cavitation. Although the process is useful in some cases, like extraction of rauwolfi a root, its large-scale application is limited due to the higher costs. One disadvantage of the procedure is the occasional but known deleterious effect of ultrasound energy (more than 20 kHz) on the active constituents of medicinal plants through formation of free radicals and consequently undesirable changes in the drug molecules.

Supercritical Fluid Extraction

Supercritical fluid extraction (SFE) is an alternative sample preparation method with general goals of reduced use of organic solvents and increased sample throughput. The factors to consider include temperature, pressure, sample volume, analyte collection, modifier (cosolvent) addition, flow and pressure control, and restrictors. Generally, cylindrical extraction vessels are used for SFE and their performance is good beyond any doubt. The collection of the extracted analyte following SFE is another important step: significant analyte loss can occur during this step, leading the analyst to believe that the actual efficiency was poor.

There are many advantages to the use of CO2 as the extracting fluid. In addition to its favorable physical properties, carbon dioxide is inexpensive, safe and abundant. But while carbon dioxide is the preferred fluid for SFE, it possesses several polarity limitations. Solvent polarity is important when extracting polar solutes and when strong analyte-matrix interactions are present. Organic solvents are frequently added to the carbon dioxide extracting fluid to alleviate the polarity limitations. Of late, instead of carbon dioxide, argon is being used because it is inexpensive and more inert. The component recovery rates generally increase with increasing pressure or temperature: the highest recovery rates in case of argon are obtained at 500 atm and 150° C. The extraction procedure possesses distinct advantages:

1. The extraction of constituents at low temperature, which strictly avoids damage from heat and some organic solvents.
2. No solvent residues.
3. Environmentally friendly extraction procedure. The largest area of growth in the development of SFE has been the rapid expansion of its applications. SFE finds extensive application in the extraction of pesticides, environmental samples, foods and fragrances, essential oils, polymers and natural products. The major deterrent in the commercial application of the extraction process is its prohibitive capital investment.

Phytonics Process

A new solvent based on hydrofl uorocarbon-134a and a new technology to optimize its remarkable properties in the extraction of plant materials offer significant environmental advantages and health and safety benefi ts over traditional processes for the production of high quality natural fragrant oils, fl avors and biological extracts. Advanced Phytonics Limited (Manchester, UK) has developed this patented technology termed "phytonics process". The products mostly extracted by this process are fragrant components of essential oils and biological or phytopharmacological extracts which can be used directly without further physical or chemical treatment.

The properties of the new generation of fluorocarbon solvents have been applied to the extraction of plant materials. The core of the solvent is 1,1,2,2-tetrafl uoroethane, better known as hydrofluorocarbon-134a (HFC-134a). This product was developed as a replacement for chlorofluorocarbons. The boiling point of this solvent is -25° C. It is not flammable or toxic. Unlike chlorofluorocarbons, it does not deplete the ozone layer. It has a vapor pressure of 5.6 bar at ambient temperature. By most standards this is a poor solvent. For example, it does not mix with mineral oils or triglycerides and it does not dissolve plant wastes. The process is advantageous in that the solvents can be customized: by using modified solvents with HFC-134a, the process can be made highly selective in extracting a specific class of phytoconstituents.

Similarly, other modified solvents can be used to extract a broader spectrum of components. The biological products made by this process have extremely low residual solvent. The residuals are invariably less than 20 parts per billion and are frequently below levels of detection. These solvents are neither acidic nor alkaline and, therefore, have only minimal potential reaction effects on the botanical materials. The processing plant is totally sealed so that the solvents are continually recycled and fully recovered at the end of each production cycle. The only utility needed to operate these systems is electricity and, even then, they do no consume much energy. There is no scope for the escape of the solvents. Even if some solvents do escape, they contain no chlorine and therefore pose no threat to the ozone layer. The waste biomass from these plants is dry and "ecofriendly" to handle.

2.1.2 Advantages of Process

- ☆ Unlike other processes that employ high temperatures, the phytonics process is cool and gentle and its products are never damaged by exposure to temperatures in excess of ambient.
- ☆ No vacuum stripping is needed which, in other processes, leads to the loss of precious volatiles.
- ☆ The process is carried out entirely at neutral pH and, in the absence of oxygen, the products never suffer acid hydrolysis damage or oxidation.
- ☆ The technique is highly selective, offering a choice of operating conditions and hence a choice of end products.
- ☆ It is less threatening to the environment.
- ☆ It requires a minimum amount of electrical energy.
- ☆ It releases no harmful emissions into the atmosphere and the resultant waste products (spent biomass) are innocuous and pose no effluent disposal problems.

☆ The solvents used in the technique are not flammable, toxic or ozone depleting.

☆ The solvents are completely recycled within the system.

2.1.3 Applications

The phytonics process can be used for extraction in biotechnology (e.g for the production of antibiotics), in the herbal drug industry, in the food, essential oil and flavor industries, and in the production of other pharmacologically active products. In particular, it is used in the production of topquality pharmaceutical-grade extracts, pharmacologically active intermediates, antibiotic extracts and phytopharmaceuticals. However, the fact that it is used in all these areas in no way prevents its use in other areas. The technique is being used in the extraction of high-quality essential oils, oleoresins, natural food colors, flavors and aromatic oils from all manner of plant materials. The technique is also used in refining crude products obtained from other extraction processes. It provides extraction without waxes or other contaminants. It helps remove many biocides from contaminated biomass.

2.1.4 Parameters for Selecting an Appropriate Extraction Method

1. Authentication of plant material should be done before performing extraction. Any foreign matter should be completely eliminated.
2. Use the right plant part and, for quality control purposes, record the age of plant and the time, season and place of collection.
3. Conditions used for drying the plant material largely depend on the nature of its chemical constituents. Hot or cold blowing air flow for drying is generally preferred. If a crude drug with high moisture content is to be used for extraction, suitable weight corrections should be incorporated.
4. Grinding methods should be specified and techniques that generate heat should be avoided as much as possible.
5. Powdered plant material should be passed through suitable sieves to get the required particles of uniform size.
6. Nature of constituents:

☆ If the therapeutic value lies in non-polar constituents, a non-polar solvent may be used. For example, lupeol is the active constituent of Crataeva nurvala and, for its extraction, hexane is generally used. Likewise, for plants like Bacopa monnieri and Centella asiatica, the active constituents are glycosides and hence a polar solvent like aqueous methanol may be used.

☆ If the constituents are thermolabile, extraction methods like cold maceration, percolation and CCE are preferred. For thermostable constituents, Soxhlet extraction (if nonaqueous solvents are used) and decoction (if water is the menstruum) are useful.

☆ Suitable precautions should be taken when dealing with constituents that degrade while being kept in organic solvents, e.g. flavonoids and phenyl propanoids.

☆ In case of hot extraction, higher than required temperature should be avoided. Some glycosides are likely to break upon continuous exposure to higher temperature.

☆ Standardization of time of extraction is important, as:

- Insufficient time means incomplete extraction.
- If the extraction time is longer, unwanted constituents may also be extracted. For example, if tea is boiled for too long, tannins are extracted which impart astringency to the fi nal preparation.

☆ The number of extractions required for complete extraction is as important as the duration of each extraction.

7. The quality of water or menstruum used should be specified and controlled.
8. Concentration and drying procedures should ensure the safety and stability of the active constituents. Drying under reduced pressure (e.g. using a Rotavapor) is widely used. Lyophilization, although expensive, is increasingly employed.
9. The design and material of fabrication of the extractor are also to be taken into consideration.
10. Analytical parameters of the final extract, such as TLC and HPLC fingerprints, should be documented to monitor the quality of different batches of the extracts.

2.1.5 Steps Involved in the Extraction of Medicinal Plants

In order to extract medicinal ingredients from plant material, the following sequential steps are involved:

1. Size reduction
2. Extraction
3. Filtration
4. Concentration

1. Size Reduction

The dried plant material is disintegrated by feeding it into a hammer mill or a disc pulverizer which has built-in sieves. The particle size is controlled by varying the speed of the rotor clearance between the hammers and the lining of the grinder and also by varying the opening of the discharge of the mill. Usually, the plant material is reduced to a size between 30 and 40 mesh, but this can be changed if the need arises.

The objective for powdering the plant material is to rupture its organ, tissue and cell structures so that its medicinal ingredients are exposed to the extraction solvent. Furthermore, size reduction maximizes the surface area, which in turn enhances the mass transfer of active principle from plant material to the solvent. The 30-40 mesh size is optimal, while smaller particles may become slimy during extraction and create difficulty during filtration.

2. Extraction

Extraction of the plant material is carried out in three ways:

1. Cold aqueous percolation
2. Hot aqueous extraction (decoction)
3. Solvent extraction (cold or hot)

Cold Aqueous Percolation

The powdered material is macerated with water and then poured into a tall column. Cold water is added until the powdered material is completely immersed. It is allowed to stand for 24 h so that water-soluble ingredients attain equilibrium in the water. The enriched aqueous extract is concentrated in multiple-effect evaporators to a particular concentration. Some diluents and excipients are added to this concentrated extract, which is then ready for medicinal use.

Hot Aqueous Extraction (Decoction)

This is done in an open-type extractor. The extractor is a cylindrical vessel made from type 316 stainless steel and has a diameter (D) greater than the height (H), i.e. the H/D ratio is approximately 0.5. The bottom of the vessel is welded to the dished end and is provided with an inside false bottom with a filter cloth. The outside vessel has a steam jacket and a discharge valve at the bottom. One part powdered plant material and sixteen parts demineralized water are fed into the extractor. Heating is done by injecting steam into the jacket. The material is allowed to boil until the volume of water is reduced to one-fourth its original volume. By this time the medicinal ingredients present in the plant material have been extracted out.

3 Filtration

The extract so obtained is separated out from the marc (exhausted plant material) by allowing it to trickle into a holding tank through the built-in false bottom of the extractor, which is covered with a fi lter cloth. The marc is retained at the false bottom, and the extract is received in the holding tank. From the holding tank, the extract is pumped into a sparkler filter to remove fi ne or colloidal particles from the extract.

Spray Drying

The filtered extract is subjected to spray drying with a high pressure pump at a controlled feed rate and temperature, to get dry powder. The desired particle size of the product is obtained by controlling the inside temperature of the chamber and by varying the pressure of the pump. The dry powder is mixed with suitable diluents or excipients and blended in a double cone mixer to obtain a homogeneous powder that can be straightaway used, for example, for filling in capsules or making tablets.

Solvent Extraction

The principle of solid-liquid extraction is that when a solid material comes in contact with a solvent, the soluble components in the solid material move to the solvent. Thus, solvent extraction of plant material results in the mass transfer of soluble active principle (medicinal ingredient) to the solvent, and this takes place in a concentration gradient. The rate of mass transfer decreases as the concentration of active principle in the solvent increases, until equilibrium is reached, i.e. the concentrations of active principle in the solid material and the solvent are the same. Thereafter, there will no longer be a mass transfer of the active principle from plant material to the solvent. Since mass transfer of the active principle also depends on its solubility in the solvent, heating the solvent can enhances the mass transfer. Moreover, if the solvent in equilibrium with the plant material is replaced with fresh solvent, the concentration gradient is changed. This gives rise to different types of extractions: cold percolation, hot percolation and concentration.

Cold Percolation

The extraction of plant material is carried out in a percolator which is a tall cylindrical vessel with a conical bottom and a built-in false bottom with a filter cloth. The percolator is connected to a condenser and a receiver for stripping solvent from the marc.

The powdered material is fed into the percolator along with a suitable solvent (ethyl alcohol or another non-polar solvent). The material is left in contact with the solvent until equilibrium of the active principle is achieved. The solvent extract, known as miscella, is taken out from the bottom discharge valve of the percolator. Fresh solvent is added into the percolator and the miscella is drained out after acquiring equilibrium. Overall, the plant material is washed four to fi ve times until it gets exhausted. All washes from the percolator are pooled and concentrated.

The solvent in the marc is stripped out by passing steam from the bottom of the percolator. The solvent and steam vapors rise and are condensed in a tubular condenser. The condensate, which is a mixture of alcohol and water, is collected in a receiver and then subjected to fractional distillation to get 95% pure ethyl alcohol which is again used as a fresh solvent. This type of percolation is not efficient as it takes a long time to reach equilibrium due to the slow mass transfer rate. The mass transfer rate can be enhanced if some sort of movement is created between the particles and the solvent.

This can be achieved either by providing inside agitation with a mechanical stirrer or by repeated circulation of the extract back to the percolator. The first method is cumbersome and power intensive whereas the latter has been successful. A circulation pump that continuously circulates the miscella back to the top of the percolator gives a better mass transfer rate and reduces the equilibrium time considerably. Still, this type of percolation is energy-consuming as large amounts of miscella from multiple washes must be concentrated to remove the solvent. To overcome this problem, a battery of percolators can be connected in series. If three washes are required for completion of the extraction, four percolators are connected in series with their respective miscellastorage tanks.

At a particular time, one percolator is out of circuit, for charging and discharging the material and also for stripping solvent from the marc, whereas the other three percolators are in operation. Material is fed into all the percolators and the solvent is fed into the first percolator. When the equilibrium in the first percolator is reached, the extract from the first percolator is sent to the second percolator. The first percolator is again filled with fresh solvent. The extract of second percolator is transferred to the third, the extract of first is transferred to second, and fresh solvent is added to the first. The extract of the third percolator is transferred to the fourth percolator.

After attaining equilibrium, the extract from the fourth percolator is drained off. The extract of the third percolator goes to fourth, the extract of second goes to third, and the extract of first goes to second percolator. The material of the first percolator, which has received three washes, is completely exhausted. This percolator is taken out of the system for stripping the solvent and discharging the extracted marc. This is again filled with fresh plant material and the sequence is repeated with other percolators. In this way, solvent of each percolator comes in contact three times with solid material and gets fully enriched with active principle. The enriched extract is sent for solvent recovery and concentration. Thus, instead of concentrating three volumes of solvent, only one volume has to be concentrated; this saves energy and the process is efficient.

Hot Percolation

Increasing the temperature of the solvent increases the solubility of the active principle, which increases the concentration gradient and therefore enhances the mass transfer of active principle from solid material to the solvent, provided the active principle is not heat sensitive. This is achieved by incorporating a heat exchanger between the circulation pump and the feed inlet of the percolator. The extract is continuously pumped into a tubular heat exchanger which is heated by steam. The temperature of the extract in the percolator is controlled by a steam solenoid valve through a temperature indicator controller. This sort of arrangement can be incorporated in single percolators or in a battery of percolators as needed.

The percolators that are tall cylindrical towers must be housed in sheds of relatively great height. Tall towers are difficult to operate, especially when charging material and discharging the marc from the top and bottom manholes, which are time-consuming and labor-intensive procedures. Tall towers have been replaced by extractors of smaller height for which the H/D ratio is not more than 1.5.

These extractors have perforated baskets in which the material to be extracted is charged. These perforated baskets, when loaded outside, can be inserted into the extractor with a chain pulley block and, after the extraction, they can be lifted out from the extractor for discharging the marc. Some extractors have an electrical hoist for the charging the material and discharging the marc, which makes the operation less labor-intensive, quick and efficient.

The other type of instrument for extraction of medicinal ingredients from plant material is the Soxhlet apparatus, which consists of an extractor, a distillation still, a tubular condenser for the distillation still, a tubular condenser for the recovery of solvent from the marc, a receiver for collecting the condensate from the condenser, and a solvent storage tank. The plant material is fed into the extractor, and solvent is added until it reaches the siphon point of the extractor. Then, the extract is siphoned out into the distillation still, which is heated with steam. The solvent vapors go to the distillation condenser, get condensed and return to the extractor. The level of the solvent in the extractor again rises to the siphon point and the extract is siphoned out into the distillation still. In this way, fresh solvent comes in contact with the plant material a number of times, until the plant material is completely extracted. The fi nal extract in the distillation still, which is rich in active principle, is concentrated and the solvent is recovered.

4 Concentration

The enriched extract from percolators or extractors, known as miscella, is fed into a wiped fi lm evaporator where it is concentrated under vacuum to produce a thick concentrated extract. The concentrated extract is further fed into a vacuum chamber dryer to produce a solid mass free from solvent. The solvent recovered from the wiped film evaporator and vacuum chamber dryer is recycled back to the percolator or extractor for the next batch of plant material. The solid mass thus obtained is pulverized and used directly for the desired pharmaceutical formulations or further processed for isolation of its phytoconstituents.

2.2. EXTRACTION OF AROMATIC PLANTS

The types of volatile isolates that are obtained commercially from aromatic plants are essential oils, concretes, absolutes, pomades and resinoids. Essential oils are isolated from plant material by distillation whereas other volatile isolates are obtained by solvent extraction.

2.2.1 Features

1 Concrete

This is an extract of fresh flowers, herbs, leaves and the flowering tops of plants obtained by the use of a hydrocarbon solvent such as butane, pentane, hexane and petroleum ether. Concrete is rich in hydrocarbonsoluble material and devoid of water-soluble components. It is generally a waxy, semisolid, dark-colored material free from the original solvent. In practice, concretes are produced in static extractors. These extractors are fi tted with numerous perforated trays so that the flowers do not get compressed by their own weight. Each perforated tray has a spacer so the number and distance between them are predetermined. The set of perforated trays can be within a removable cylindrical basket. In the centre of the lower tray, there is a rod on which the spacers and the perforated trays are fitted while at the top there is a ring or a hook so that the entire contents of the extractor can be readily removed by a chain pulley block.

While stacking the flowers on these trays, care should be taken to minimize bruising and damage of the flowers, because such damage can result in the release of enzymes in the flower juice which deteriorates the quality of concrete. The basket stacked with flowers is inserted into the extractor and the solvent of choice is introduced from the bottom into the extractor until the material on the perforated disc assembly is completely immersed. Four to five such washes are given until the material is exhausted.

The enriched solvent from the extractor is pumped into an evaporator for solvent recovery and the solvent content is reduced to about one-tenth the original volume. The recovered solvent is pumped to the solvent tanks to be used again. The concentrated material from the evaporator is pumped into a vacuum evaporator where the solvent is removed more carefully under high vacuum and the recovered solvent is returned to the solvent tanks for repeated use. The resultant concrete has an odor similar to but stronger than the material from which it was extracted. In concrete manufacturing, it is a normal practice to circulate fresh solvent through a battery of extractors. At each cycle, the solvent becomes more enriched with the flower volatiles until extraction is complete. The number of extractors has to be synchronized with the number of solvent washes.

2 Absolutes

Concretes are not widely used in perfumery in their native form but are generally converted into an alcohol-soluble volatile concentrate known as an absolute, i.e. they have to be extracted with alcohol.

To make an absolute, the concrete is mixed with absolute alcohol and agitated thoroughly in a vessel with an agitator. During agitation, the temperature is kept at 40°-60° C and the concrete is immersed in the solution. The solution is cooled down to -5° to -10° C to precipitate out the wax, since waxes are normally insoluble in alcohol below -1° C. The precipitated wax is removed by passing the solution through a rotary filter. The filtrate from the rotary filter is pumped into a primary evaporator, where it is concentrated to about 10% alcohol content. Finally, the concentrated extract is pumped into an agitating-type evaporator, where the alcohol is carefully removed under high vacuum.

3 Resinoids

Resinoid is an extract of naturally resinous material, made with a hydrocarbon solvent. Resinoids are usually obtained from dry materials. The extraction process is same as that of concrete production, except that perforated discs are not used for stacking the material; instead powder from dry plant material is fed into the extractor.

4 Pomades

Pomades are obtained by a process known as enfleurage, which is a cold fat extraction method. The fat is spread out on glass plates contained in wooden frames, leaving a clear margin near the edges. The absorptive surface of the fat is increased by surface grooves made with a wooden spatula.

Fresh flowers are spread out on the surface of the fat and the frames are stacked in piles. After the perfume oils have been absorbed from the flowers, the spent flowers are removed by hand. Fresh flowers are again spread on the fat surface. This is repeated until the fat surface is completely enriched with perfume oils. The pomade so obtained is ready for cold alcoholic extraction.

5 Essential Oils

Essential oils are used in a wide variety of consumer goods such as detergents, soaps, toilet products, cosmetics, pharmaceuticals, perfumes, confectionery food products, soft drinks, distilled alcoholic beverages (hard drinks) and insecticides. The world production and consumption of essential oils and perfumes are increasing very fast. Production technology is an essential element to improve the overall yield and quality of essential oil. The traditional technologies pertaining to essential oil processing are of great significance and are still being used in many parts of the globe. Water distillation, water and steam distillation, steam distillation, cohobation, maceration and enfleurage are the most traditional and commonly used methods. Maceration is adaptable when oil yield from distillation is poor. Distillation methods are good for powdered almonds, rose petals and rose blossoms, whereas solvent extraction is suitable for expensive, delicate and thermally unstable materials like jasmine, tuberose, and hyacinth. Water distillation is the most favored method of production of citronella oil from plant material.

2.2.2 Sources of Natural Essential Oils

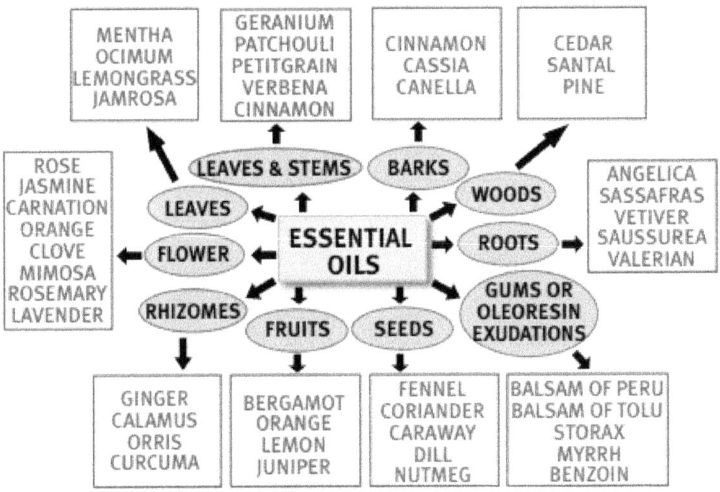

Plant organs containing essential oils

Plant organs containing natural essential oils are illustrated in Figure. Essential oils are generally derived from one or more plant parts, such as flowers (e.g. rose, jasmine, carnation, clove, mimosa, rosemary, lavander), leaves (e.g. mint, Ocimum spp., lemongrass, jamrosa), leaves and stems (e.g. geranium, patchouli, petitgrain, verbena, cinnamon), bark (e.g. cinnamon, cassia, canella), wood (e.g. cedar, sandal, pine), roots (e.g. angelica, sassafras, vetiver, saussurea, valerian), seeds (e.g fennel, coriander, caraway, dill, nutmeg), fruits (bergamot, orange, lemon, juniper), rhizomes (e.g. ginger, calamus, curcuma, orris) and gums or oleoresin exudations (e.g. balsam of Peru, balsam of Tolu, storax, myrrh, benzoin).

Specialized plant structures that produce and store essential oils are shown in Figure. Depending upon the plant family, essential oils may occur in specialized secretary structures such as glandular hairs (Labiatae, Verbenaceace, Geraniaceae), modifi ed parenchymal cells (Piperaceae), resin canals (conifers), oil tubes called vittae (Umbelliferae), lysigenous cavities (Rutaceae), schizogenous passages (Myrtaceae, Graminae, Compositae) or gum canals (Cistacae, Burseraceae). It is well known that when a geranium leaf is lightly touched, an odor is emitted because the long stalked oil glands are fragile. Similarly, the application of slight pressure on a peppermint leaf will rupture the oil gland and release oil. In contrast, pine needles and eucalyptus leaves do not release their oils until the epidermis of the leaf is broken. Hence, the types of structures in which oil is contained differ depending on the plant type

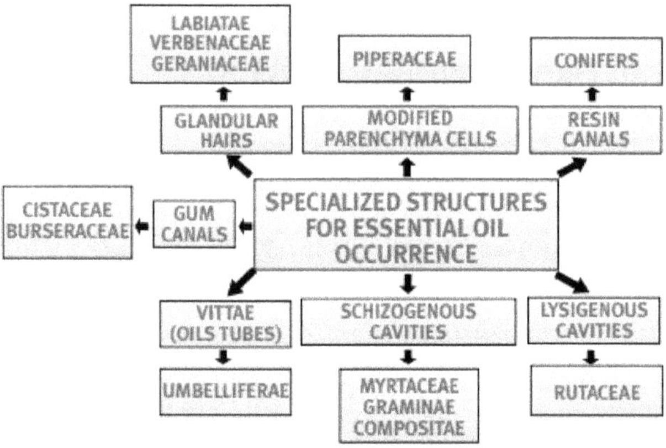

Family-specific plant tissues responsible for producing or storing essential oil

During handling, some flowers continue to produce aroma while other quickly loose their odor. Flowers collected at different times may also give different perfumery values. Regarding the rose, half-open flowers with plump anthers give higher oil yield than fully opened flowers with shrivelled anthers. Humidity, wind, rain and surface temperature also affect the oil yield considerably. Harvesting schedule affects both quantity and quality of the oil.

2.2.3 Essential Oil Constituents

Major constituents of essential oils are shown in Figure 5, from which it is clear that most essential oils consist of hydrocarbons, esters, terpenes, lactones, phenols, aldehydes, acids, alcohols, ketones, and esters. Among these, the oxygenated compounds (alcohols, esters, aldehydes, ketones, lactones, phenols) are the principal odor source. They are more stable against oxidizing and resinifying influences than other constituents. On the other hand, unsaturated constituents like monoterpenes and sesquiterpenes have the tendency to oxidize or resinify in the presence of air and light. The knowledge of individual constituents and their physical characteristics, such as boiling point, thermal stability and vapor-pressure-temperature relationship, is of paramount importance in technology development of oxygenated compounds.

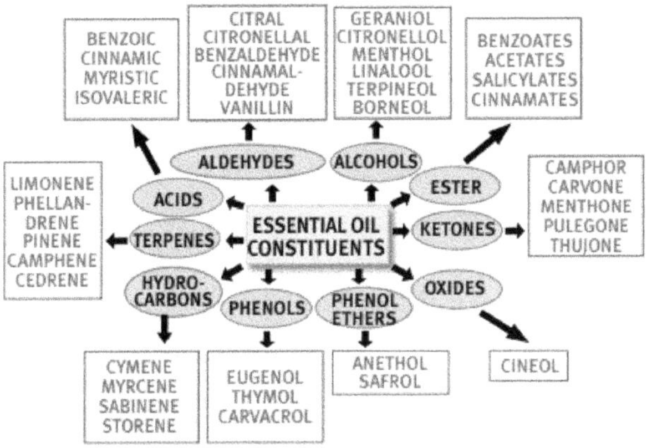

Heterogeneous chemical groups present in essential oil

2.2.4 Methods of Producing Essential Oils

Methods for producing essential oils from plant materials are summarized in Figure 6. Regarding hydrodistillation, the essential oils industry has developed terminology to distinguish three types:

1. water distillation
2. water and steam distillation
3. direct steam distillation

Originally introduced by Von Rechenberg, these terms have become established in the essential oil industry. All three methods are subject to the same theoretical considerations which deal with distillation of two-phase systems. The differences lie mainly in the methods of handling the material.

Some volatile oils cannot be distilled without decomposition and thus are usually obtained by expression (lemon oil, orange oil) or by other mechanical means. In certain countries, the general method for obtaining citrus oil involves puncturing the oil glands by rolling the fruit over a trough lined with sharp projections that are long enough to penetrate the epidermis and pierce the oil glands located within outer portion of the peel (ecuelle method). A pressing action on the fruit removes the oil from the glands, and a fi ne spray of water washes the oil from the mashed peel while the juice is extracted through a central tube that cores the fruit. The resulting oil-water emulsion is separated by centrifugation. A variation of this process is to remove the peel from the fruit before the oil is extracted.

Often, the volatile oil content of fresh plant parts (flower petals) is so small that oil removal is not commercially feasible by the aforementioned methods. In such instances, an odorless, bland, fi xed oil or fat is spread in a thin layer on glass plates. The flower petals are placed on the fat for a few hours; then repeatedly, the oil petals are removed, and a new layer of petals is introduced. After the fat has absorbed as much fragrance as possible, the oil may be removed by extraction with alcohol. This process, known as enfl eurage, was formerly used extensively in the production of perfumes and pomades.

In the perfume industry, most modern essential oil production is accomplished by extraction, using volatile solvents such as petroleum ether and hexane. The chief advantages of extraction over distillation is that uniform temperature (usually 50° C) can be maintained during the process, As a result, extracted oils have a more natural odor that is unmatched by distilled oils, which may have undergone chemical alteration by the high temperature. This feature is of considerable importance to the perfume industry; however, the established distillation method is of lower cost than the extraction process.

Destructive distillation means distilling volatile oil in the absence of air. When wood or resin of members of the Pinaceae or Cupressaceae is heated without air, decomposition takes place and a number of volatile compounds are driven off. The residual mass is charcoal. The condensed volatile matter usually separates into 2 layers: an aqueous layer containing wood naptha (methyl alcohol) and pyroligneous acid (crude acetic), and a tarry liquid in the form of pine tar, juniper tar, or other tars, depending on the wood used. This dry distillation is usually conducted in retorts and, if the wood is chipped or coarsely ground and the heat is applied rapidly, the yield often represents about 10% of the wood weight used.

Hydrodistillation

In order to isolate essential oils by hydrodistillation, the aromatic plant material is packed in a still and a suffi cient quantity of water is added and brought to a boil; alternatively, live steam is injected into the plant charge. Due to the infl uence of hot water and steam, the essential oil is freed from the oil glands in the plant tissue. The vapor mixture of water and oil is condensed by indirect cooling with water. From the condenser, distillate fl ows into a separator, where oil separates automatically from the distillate water.

Mechanism of Distillation

Hydrodistillation of plant material involves the following main physicochemical processes:

1. Hydrodiffusion

2. Hydrolysis
3. Decomposition by heat

Hydrodiffusion

Diffusion of essential oils and hot water through plant membranes is known as hydrodiffusion. In steam distillation, the steam does not actually penetrate the dry cell membranes. Therefore, dry plant material can be exhausted with dry steam only when all the volatile oil has been freed from the oil-bearing cells by first thorough comminution of the plant material.

But, when the plant material is soaked with water, exchange of vapors within the tissue is based on their permeability while in swollen condition. Membranes of plant cells are almost impermeable to volatile oils. Therefore, in the actual process, at the temperature of boiling water, a part of volatile oil dissolves in the water present within the glands, and this oil-water solution permeates, by osmosis, the swollen membranes and fi nally reaches the outer surface, where the oil is vaporized by passing steam.

Another aspect of hydrodiffusion is that the speed of oil vaporization is not infl uenced by the volatility of the oil components, but by their degree of solubility in water. Therefore, the high-boiling but more water-soluble constituents of oil in plant tissue distill before the low-boiling but less water-soluble constituents. Since hydrodiffusion rates are slow, distillation of uncomminuted material takes longer time than comminuted material.

Hydrolysis

Hydrolysis in the present context is defi ned as a chemical reaction between water and certain constituents of essential oils. Esters are constituents of essential oils and, in the presence of water, especially at high temperatures, they tend to react with water to form acids and alcohols.

Effect of Heat

Almost all constituents of essential oils are unstable at high temperature. To obtain the best quality oil, distillation must be done at low temperatures. The temperature in steam distillation is determined entirely by the operating pressure, whereas in water distillation and in water and steam distillation the operating pressure is usually atmospheric.

All the previously described three effects, i.e. hydrodiffusion, hydrolysis and thermal decomposition, occur simultaneously and affect one another. The rate of diffusion usually increases with temperatures as does the solubility of essential oils in water. The same is true for the rate and extent of hydrolysis. However, it is possible to obtain better yield and quality of oils by:

1. maintaining the temperature as low as possible
2. using as little water as possible, in the case of steam distillation
3. thoroughly comminuting the plant material and packing it uniformly before distillation.

Three Types of Hydrodistillation

Three are three types of hydrodistillation for isolating essential oils from plant materials:

1. Water distillation
2. Water and steam distillation
3. Direct steam distillation

Water Distillation

In this method, the material is completely immersed in water, which is boiled by applying heat by direct fi re, steam jacket, closed steam jacket, closed steam coil or open steam coil. The main characteristic of this process is that there is direct contact between boiling water and plant material.

When the still is heated by direct fi re, adequate precautions are necessary to prevent the charge from overheating. When a steam jacket or closed steam coil is used, there is less danger of overheating; with open steam coils this danger is avoided. But with open steam, care must be taken to prevent accumulation of condensed water within the still. Therefore, the still should be well insulated. The plant material in the still must be agitated as the water boils, otherwise agglomerations of dense material will settle on the bottom and become thermally degraded. Certain plant materials like cinnamon bark, which are rich in mucilage, must be powdered so that the charge can readily disperse in the water; as the temperature of the water increases, the mucilage will be leached from the ground cinnamon. This greatly increases the viscosity of the water-charge mixture, thereby allowing it to char. Consequently, before any field distillation is done, a small-scale water distillation in glassware should be performed to observe whether any changes take place during the distillation process. From this laboratory trial, the yield of oil from a known weight of the plant material can be determined.

The laboratory apparatus recommended for trial distillations is the Clevenger system. During water distillation, all parts of the plant charge must be kept in motion by boiling water; this is possible when the distillation material is charged loosely and remains loose in the boiling water. For this reason only, water distillation possesses one distinct advantage, i.e. that it permits processing of fi nely powdered material or plant parts that, by contact with live steam, would otherwise form lumps through which the steam cannot penetrate. Other practical advantages of water distillation are that the stills are inexpensive, easy to construct and suitable for field operation. These are still widely used with portable equipment in many countries.

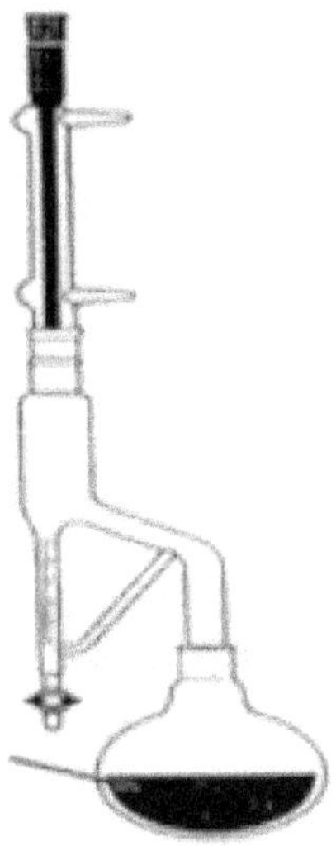

Clevenger-type laboratory-scale hydrodistillation apparatus

The main disadvantage of water distillation is that complete extraction is not possible. Besides, certain esters are partly hydrolyzed and sensitive substances like aldehydes tend to polymerize. Water distillation requires a greater number of stills, more space and more fuel. It demands considerable experience and familiarity with the method. The high-boiling and somewhat water-soluble oil constituents cannot be completely vaporized or they require large quantities of steam. Thus, the process becomes uneconomical. For these reasons, water distillation is used only in cases in which the plant material by its very nature cannot be processed by water and steam distillation or by direct steam distillation.

2.2.5 Traditional Method of Producing Attar Using Hydrodistillation

Floral attars are defined as the distillates obtained by hydrodistillation of flowers (such as saffron, marigold, rose, jasmine, pandanus) in sandal wood oil or other base materials like paraffin. Attar manufacturing takes place in remote places because the flowers must be processed quickly after collection. The apparatus and equipment used to manufacture attar are light, flexible, easy to repair, and have a fair degree of efficiency. Keeping in view these facts, the traditional "deg and bhapka" process has been used for centuries and is used even now with the following traditional equipment.

- ☆ Deg (still)
- ☆ Bhapka (receiver)
- ☆ Chonga (bamboo condenser)
- ☆ Traditional bhatti (furnace)
- ☆ Gachchi (cooling water tank)
- ☆ Kuppi (leather bottle)

Water and Steam Distillation

In water and steam distillation, the steam can be generated either in a satellite boiler or within the still, although separated from the plant material. Like water distillation, water and steam distillation is widely used in rural areas. Moreover, it does not require a great deal more capital expenditure than water distillation. Also, the equipment used is generally similar to that used in water distillation, but the plant material is supported above the boiling water on a perforated grid. In fact, it is common that persons performing water distillation eventually progress to water and steam distillation.

It follows that once rural distillers have produced a few batches of oil by water distillation, they realize that the quality of oil is not very good because of its still notes (subdued aroma). As a result, some modifications are made. Using the same still, a perforated grid or plate is fashioned so that the plant material is raised above the water. This reduces the capacity of the still but affords a material quite wet. This slows down distillation as the steam has to vaporize the water to allow it to condense further up the still. One way to prevent the lower plant material resting on the grid from becoming waterlogged is to use a baffle to prevent the water from boiling too vigorously and coming in direct contact with the plant material.

1. Advantages of Water and Steam Distillation over Water Distillation

- ☆ Higher oil yield.
- ☆ Components of the volatile oil are less susceptible to hydrolysis and polymerization (the control of wetness on the bottom of the still affects hydrolysis, whereas the thermal conductivity of the still walls affects polymerization).
- ☆ If refluxing is controlled, then the loss of polar compounds is minimized.
- ☆ Oil quality produced by steam and water distillation is more reproducible.
- ☆ Steam and water distillation is faster than water distillation, so it is more energy efficient. Many oils are currently produced by steam and water

distillation, for example lemongrass is produced in Bhutan with a rural steam and water distillation system.

2. Disadvantages of Water and Steam Distillation

- ☆ Due to the low pressure of rising steam, oils of high-boiling range require a greater quantity of steam for vaporization - hence longer hours of distillation.
- ☆ The plant material becomes wet, which slows down distillation as the steam has to vaporize the water to allow it to condense further up the still.
- ☆ To avoid that the lower plant material resting on the grid becomes waterlogged, a baffl e is used to prevent the water from boiling too vigorously and coming in direct contact with the plant material.

Direct Steam Distillation

As the name suggests, direct steam distillation is the process of distilling plant material with steam generated outside the still in a satellite steam generator generally referred to as a boiler. As in water and steam distillation, the plant material is supported on a perforated grid above the steam inlet. A real advantage of satellite steam generation is that the amount of steam can be readily controlled. Because steam is generated in a satellite boiler, the plant material is heated no higher than 100° C and, consequently, it should not undergo thermal degradation. Steam distillation is the most widely accepted process for the production of essential oils on large scale. Throughout the flavor and fragrance supply business, it is a standard practice.

An obvious drawback to steam distillation is the much higher capital expenditure needed to build such a facility. In some situations, such as the large-scale production of low-cost oils (e.g. rosemary, Chinese cedarwood, lemongrass, litsea cubeba, spike lavender, eucalyptus, citronella, cornmint), the world market prices of the oils are barely high enough to justify their production by steam distillation without amortizing the capital expenditure required to build the facility over a period of 10 years or more.

1 Advantages of Direct Steam Distillation

- ☆ Amount of steam can be readily controlled.
- ☆ No thermal decomposition of oil constituents.
- ☆ Most widely accepted process for large-scale oil production, superior to the other two processes.

2 Disadvantage of Direct Steam Distillation

- ☆ Much higher capital expenditure needed to establish this activity than for the other two processes.

2.2.6 Essential Oil Extraction by Hydrolytic Maceration Distillation

Certain plant materials require maceration in warm water before they release their essential oils, as their volatile components are glycosidically bound. For example, leaves of wintergreen (Gaultheria procumbens) contain the precursor gaultherin and the enzyme primeverosidase; when the leaves are macerated in warm water, the enzyme acts on the gaultherin and liberates free methyl salicylate and primeverose. Other similar examples include brown mustard (sinigrin), bitter almonds (amygdalin) and garlic (alliin).

1. Essential Oil Extraction by Expression

Expression or cold pressing, as it is also known, is only used in the production of citrus oils. The term expression refers to any physical process in which the essential oil glands in the peel are crushed or broken to release the oil. One method that was practiced many years ago, particularly in Sicily (spugna method), commenced with halving the citrus fruit followed by pulp removal with the aid of sharpened spoon-knife (known as a rastrello). The oil was removed from the peel either by pressing the peel against a hard object of baked clay (concolina) which was placed under a large natural sponge or by bending the peel into the sponge. The oil emulsion absorbed by the sponge was removed by squeezing it into the concolina or some other container. It is reported that oil produced this way contains more of the fruit odor character than oil produced by any other method.

A second method known as equaling (or the scodella method), uses a shallow bowl of copper (or sometimes brass) with a hollow central tube; the equaling tool is similar in shape to a shallow funnel. The bowl is equipped with brass points with blunt ends across which the whole citrus fruit is rolled by hand with some pressure until all of the oil glands have burst. The oil and aqueous cell contents are allowed to dribble down the hollow tube into a container from which the oil is separated by decantation. Obviously, hand pressing is impractical because it is an extremely slow process, e.g. on average only 2-4 lbs oil per day can be produced by a single person using one of these hand methods. As a result, over the years a number of machines have been designed to either crush the peel of a citrus fruit or crush the whole fruit and then separate the oil from the juice.

2. Pelatrice Process

In the pelatrice process, citrus fruits are fed from a hopper into the abrasive shell of the machine. The fruits are rotated against the abrasive shell by a slow-moving Archimedian screw whose surface rasps the fruit surfaces causing some of the essential oil cavities on the peel to burst and release their oil-water emulsion. This screw further transports the fruit into a hopper in which rollers covered with abrasive spikes burst the remaining oil cavities. The oil and water emulsion is washed away from the fruit by a fine spray of water. The emulsion next passes through a separator where any solids are removed, after which it passes through two centrifugal separators working in series to yield the pure oil. Most bergamot oil and some lemon oil are produced this way in Italy.

3.Sfumatrice Process

The sfumatrice equipment consists of a metallic chain that is drawn by two horizontal ribbed rollers. The peels are conveyed through these rollers during which time they are pressed and bent to release their oil. As in pelatrice, the oil is washed away from the sfumatrice rollers by fi ne sprays of water. Again, the oil is initially passed through a separator prior to being sent to two centrifuges in series, so that purified oil can be produced. At one time, sfumatrice was the most popular process for citrus oil isolation in Italy; however, today the pelatrice method appears more popular.

Essential Oil Extraction with Cold Fat (Enfleurage)

Despite the introduction of the modern process of extraction with volatile solvents, the old fashioned method of enfleurage, as passed on from father to son and perfected in the course of generations, still plays an important role. Enfleurage on a large scale is today carried out only in the Grasse region of France, with the possible exception of isolated instances in India where the process has remained primitive.

The principles of enfleurage are simple. Certain flowers (e.g. tuberose and jasmine) continue the physiological activities of developing and giving off perfume even after picking. Every jasmine and tuberose flower resembles, so to speak, a tiny factory continually emitting minute quantities of perfume. Fat possesses a high power of absorption and, when brought in contact with fragrant flowers, readily absorbs the perfume emitted. This principle, methodically applied on a large scale, constitutes enfleurage. During the entire period of harvest, which lasts for eight to ten weeks, batches of freshly picked flowers are strewn over the surface of a specially prepared fat base (corps), let there (for 24 h in the case of jasmine and longer in the case of tuberose), and then replaced by fresh flowers. At the end of the harvest, the fat, which is not renewed during the process, is saturated with flower oil. Thereafter, the oil is extracted from the fat with alcohol and then isolated. The success of enfleurage depends to a great extent upon the quality of the fat base employed. Utmost care must be exercised when preparing the corps. It must be practically odorless and of proper consistency.

If the corps is too hard, the blossoms will not have sufficient contact with the fat, curtailing its power of absorption and resulting in a subnormal yield of flower oil. On the other, if it is too soft, it will tend to engulf the flowers and the exhausted ones will adhere; when removed, the fl owers will retain adhering fat, resulting in considerable shrinkage and loss of corps. The consistency of the corps must, therefore, be such that it offers a semihard surface from which the exhausted flowers can easily be removed. The process of enfleurage is carried out in cool cellars, and every manufacturer must prepare the corps according to the prevailing temperature in the cellars during the months of the flower harvest. Many years of experience have proved that a mixture of one part of highly purified tallow and two parts of lard is eminently suitable for enfleurage. This mixture assures a suitable consistency of the corps in conjunction with high power of absorption. The fat corps thus prepared is white, smooth, absolutely of uniform consistency, free of water and practically odorless. Some manufacturers also add small quantities of orange flower or rose water when preparing the corps. This seems to be done for the sake of convention. Such additions somewhat shade the odor of the finished product by imparting a slight orange blossom or rose note.

Enfleurage and Defleurage

Every enfleurage building is equipped with thousands of socalled chassis, which serve as vehicles for holding the fat corps during the process. A chassis consists of a rectangular wooden frame. The frame holds a glass plate upon both sides of which the fat corps is applied with a spatula at the beginning of the enfleurage process. When piled one above the other, the chassis form airtight compartments, with a layer of fat on the upper and lower side of each glass plate.

Every morning during the harvest the freshly picked flowers arrive, and after being cleaned of impurities, such as leaves and stalks, are strewn by hand on top of the fat layer of each glass plate. Blossoms wet from dew or rain must never be employed, as any trace of moisture will turn the corps rancid. The chassis are then piled up and left in the cellars for 24 h or longer, depending upon the type of flowers. The latter rest in direct contact with one fat layer (the lower one), which acts as a direct solvent whereas the other fat layer (beneath the glass plate of the chassis above) absorbs only the volatile perfume given off by the flowers.

After 24 h, the flowers have emitted most of their oil and start to wither, developing an objectionable odor. They must then be removed from the corps, which process, despite all efforts to introduce labor-saving devices, is still done by hand. Careful removal of the flower (defleurage) is almost more important than charging the corps on the chassis with fresh flowers (enfleurage) and, therefore, the persons doing this work must be experienced and skilled. Most of the exhausted flowers will fall from the fat layer on the chassis glass plate when the chassis is struck lightly against the working table, but since it is necessary to remove every single flower and every particle of the flower, tweezers are used for this delicate operation.

Immediately following defleurage, that is, every 24 h, the chassis are recharged with fresh flowers. For this purpose the chassis are turned over and the fat layer, which in the previous operation formed the top (ceiling) of the small chamber, is now directly charged with flowers. In the case of jasmine, the entire enfleurage process lasts about 70 days: daily the exhausted flowers are removed and the chassis are recharged with fresh ones. At the beginning of, and several times during, the harvest, the fat on the chassis is scratched over with metal combs and tiny furrows are drawn in order change and increase the surface of absorption.

At the end of the harvest, the fat is relatively saturated with flower oil and possesses the typical fragrance. The perfumed fat must then be removed from the glass plates between the chassis. For this purpose, it is scraped off with a spatula and then carefully melted and bulked in closed containers. The final product is called pomade (pomade de jasmine, pomade de tuberous, pomade de violet, etc.). The most highly saturated pomade is pomade no. 36, because the corps on the chassis have been treated with fresh flowers 36 times during the whole process of enfleurage. At the beginning of the harvest, every chassis is charged with about 360 g fat corps on each side of the glass plate, in other words, with 720 g per chassis. Every kilogram of fat corps should be in contact with about 2.5 kg (preferably with 3.0 kg) of jasmine fl owers for the entire period of enfleurage, which lasts from 8 to 10 weeks. The quantities differ somewhat for different flowers. At the end of enfl eurage, the fat corps has lost about 10% of its weight because of the various manipulations.

Hot Maceration Process

In this process, the long enfleurage time is reduced by the immersion of petals in molten fat heated at 45°-60° C for 1 to 2 h, depending upon the plant species. After each immersion, the fat is filtered and separated from the petals. After 10 to 20 immersions, the fat is separated from waste fl owers and water. Absolute of maceration is then produced from fat containing oil through the process of extraction and concentration under reduced pressure. It is mainly used for highly delicate flowers whose physiological activities are lost rapidly after their harvest, such as lily of valley.

Modern (Non-traditional) Methods of Extraction of Essential Oils

Traditional methods of extraction of essential oils have been discussed and these are the methods most widely used on commercial scale. However, with technological advancement, new techniques have been developed which may not necessarily be widely used for commercial production of essential oils but are considered valuable in certain situations, such as the production of costly essential oils in a natural state without any alteration of their thermosensitive components or the extraction of essential oils for micro-analysis. These techniques are as follows:

- ☆ Headspace trapping techniques
  - Static headspace technique
  - Vacuum headspace technique
  - Dynamic headspace technique
- ☆ Solid phase micro-extraction (SPME)
- ☆ Supercritical fl uid extraction (SFE)
- ☆ Phytosol (phytol) extraction
- ☆ Protoplast technique
- ☆ Simultaneous distillation extraction (SDE)
- ☆ Microwave distillation
- ☆ Controlled instantaneous decomposition (CID)
- ☆ Thermomicrodistillation
- ☆ Microdistillation
- ☆ Molecular spinning band distillation

2.3 BIOMASS OF MEDICINAL AND AROMATIC PLANTS

Natural products from plants are well known in old world civilizations. International markets have been utilizing products from herbs since several decades. India is one of the earliest civilizations that have recognized the importance of herbal products for disease management, nutrition and beauty enhancement. With the discovery of several new molecules from herbs for treating dreaded diseases like cancer etc. and the relative safety of these products, the global demand for medicinal plant products has increased in recent years. Similarly consumers are preferring cosmetics with aromatic products from plants resulting in higher demand for the raw material. India is sitting on a treasure of 8000 medicinal and 2500 aromatic plants. In the last 50 years, Central and State Government research organizations have developed several technologies for utilizing medicinal and aromatic plants. Small and medium enterprises have already been set up by many entrepreneurs and opportunities exist for new enterprises in this field.

The primary raw material for plant based products is the biomass of the medicinal and aromatic plants. Biomass is the green or dried plant part(s) or whole plant obtained from plants collected from their natural habitats or harvested from cultivated plants. The biomass needed for the industry should pass the following criteria:

1. should be from specific plant part(s) derived from a single botanically designated species and should not be mixed with biomass from other related or unrelated plant species
2. should be free from extraneous abiotic matter like dust, stones, soil, and biotic matter such as fungus, droppings from rodents/ insects/ birds etc.
3. should be free from damage due to insects/ diseases
4. should contain the minimum specified or accepted concentration of the chemical (s)of commercial value
5. should be free from chemical contaminants such as heavy metals, pesticide residues etc.
6. should meet the specifications agreed between the supplier and the industry such as colour, admixture with other plant parts of the same plant, level and method of drying, packing etc. Reasonably priced, good quality (meeting the above criteria) biomass is the basic raw material for the traditional (Ayurveda, Siddha, Naturopathy, Tribal medicines etc.) as well as modern industries and a country that has access to such standard biomass will be at an advantage in the international market.

2.4 GLOBAL MARKET FOR MEDICINAL PLANTS AND THEIR PRODUCTS

The following table provides an idea of the global trade in medicinal plants and their products.

Table l. Global trade in medicinal plants and their products

Particulars	Trade
Medicinal plants traded in the world	400 species (75–100 in large quantities)
Value of plant based drugs	US$ 170 billion 2050: US$ 005 trillion (World Health Organization)
Herbal teas, drug adjuncts etc.	US$ 10 billion
Nutraceuticals	US$ 80 billion
Functional foods	US$ 57 billion

The above table indicates the potential of global trade based on medicinal plants. China is ahead of India in this trade and India's market share is meager providing excellent scope and opportunity for new enterprises. The major drawbacks for global positioning of Indian medicinal plant products are: dearth of scientific validity of claimed medicinal properties, quality inconsistencies and high prices of the products. Other factors include adulteration, unethical means adopted by some companies, contaminants exceeding prescribed levels, not adhering to time schedules etc. Government of India is encouraging setting up of industries and is providing training, incentives etc. and is negotiating with many countries concerning export/ import regulations etc.

2.5 GLOBAL MARKET FOR AROMATIC PLANTS AND THEIR PRODUCTS

The following table provides an idea of the global trade in aromatic plants and their products.

Particulars	Trade
Aromatic plants traded in the world	250 types of aromatic oils
Value of aromatic oils traded in the world	US$ 1.2 billion
Global trade for flavours & fragrances	US$ 20 billion
Global trade in cosmetics	US$ 150 billion

The above table indicates the potential of global trade based on aromatic plants. India is in leading position for volume and value of aromatic oils produced and traded. Indian aromatic oils are being imported by all industrially developed countries. Menthol mint (an important aromatic-cum-medicinal plant) industry in north India is a standing example of how scientific farming coupled with entrepreneurship can lead to success at national and international platforms. Ample opportunities exist to emulate this model and apply it to several other Indian medicinal and aromatic plants.

2.6 BUSINESS OPPORTUNITIES IN MEDICINAL AND AROMATIC PLANTS

Medicinal and aromatic plants are generally regarded as industrial crops as their products are used either in pharmaceutical or in flavour and fragrance industries. With increasing demand for these plants and their products, a number of business opportunities arose that can solve unemployment problem in rural areas, improve rural living standards and enable India to become an important exporter of these plants and their products. Export of value added products would fetch higher profits than export of raw material.

Table 3. Business opportunities in medicinal and aromatic plants

Particulars	Opportunities
Cultivation	Rural employment and livelihood improvements
Processing	Establishment of chilling/ distillation units, herbal extraction units, fractionating columns etc. leading to employment generation
Bye-products based industries	Manufacturing of vermicompost, perfumed waters, botanical pesticides etc.
Product development	Manufacturing of soaps, candles, ittars, perfumes, agarbatties, traditional drugs etc.
Marketing	Marketing of seeds, plants, oil, drugs, products etc.
Consultancy	Cultivation, processing, financing, marketing etc.
Quality control	Testing laboratories for quality certification
Packaging, storing	G.I. drums, bags, cartons, godowns etc.
Transport	Tractors, trucks etc.

2.7 MARKETING

The most important aspect of medicinal and aromatic plants cultivation is the of produce by the growers. The existing market scenario is depicted in figure.

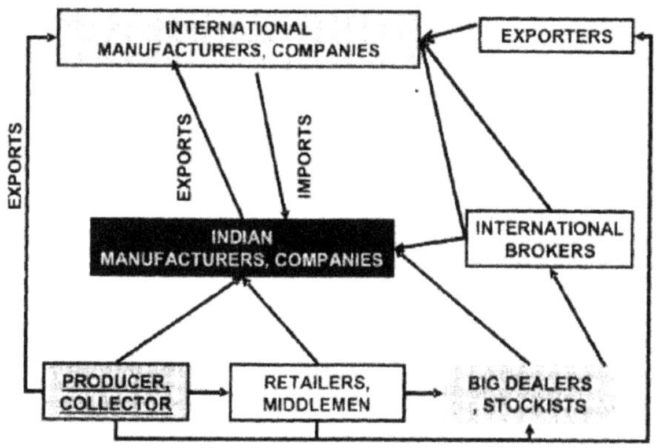

Medicinal and aromatic plants market network marketing

Producers feign ignorance concerning demand-supply position and the prevailing market prices. The prices of the commodities may fluctuate within the season resulting in lower or higher prices. Other problems in marketing are:

1. Concentration of markets at few locations
2. Absence of market networks in many States.

3. Unethical means adopted by market forces
4. Absence of minimum purchase prices
5. Absence of market yards for trading and go-downs for safe storing of produce
6. Absence of growers' organizations.

2.8 FUTURE OPPORTUNITIES IN MEDICINAL AND AROMATIC PLANTS

During the past 30 years, medicinal and aromatic plants in the United States have moved from essentially unknown, minor agricultural plants into crops that many farmers consider producing as an alternative to usual plantings of food and feed crops. The attraction of medicinal and aromatic plants as worthy farm crops has grown due to the demand created by consumer interest in these plants for culinary, medicinal, and other anthropogenic applications. As racial diversity in the US has expanded, immigrants from countries in which herbs and herbal medicines are commonly used to flavor foods and treat illnesses have introduced other Americans to a diverse range of plant materials. Indeed, market trend surveys indicate that mainstream American consumers will purchase 75% of the ethnic foods during the next decade.

Farmers growing medicinal and aromatic plants, similar to farmers in other agricultural systems, begin each growing season with hopes for success in producing a crop that brought to market will more than repay the expense of production. Yet, in addition to traditional cropping uncertainties of weather, pests, and other limitations, the medicinal and aromatic plant farmer also faces changes in consumer interest, international trade policies, and other issues that control demand. For these reasons, an understanding of future opportunities in the medicinal and aromatic plant industry is necessary to enable US growers to envision and invest in medicinal and aromatic crops that will meet market demands.

2.8.1 Perspectives

The initiation of medicinal plant and aromatic production, as a gathering or cultivation of plant materials, is lost to history, but most likely began at or near the time of the first afflictions and the recognition that smelling, chewing, and/or eating some plant materials could provide relief from nausea, pain, and/or other infirmities. Those plants containing the unique chemical profiles that offered pain relief, pleasant aromas, and enhanced food flavors would soon be renowned and much valued by early humans, leading to associations among certain aliments, plants, and "feeling better". Thus, these plants, now known as medicinal and aromatic plants, and their extracts became the main source for medicines, seasonings, colorings, preservatives, and other similar items used in societies, sustained by myths and traditions developed to explain the almost "magical" powers of selected species and to transmit the accumulation of acquired knowledge about these species before the era of written records.

As continued experimentation with various plant materials demonstrated the benefits of having specific plants immediately available for use in medical treatment and food flavoring, husbandry of these plant species undoubtedly started. Although the collection of plants probably remained the primary source of medicinal and aromatic plant material for a considerable period of time, cultivation and growth of plants could be expected to have begun in small garden plots and botanical collections. As human migration led to settlements within various ecosystems, species having medicinal and aromatic properties specific to those regions would be discovered, leading to a collection of plant materials with a variety of uses and the initiation of trade among neighboring groups for unavailable plant materials. This initial exchange of plant material could be expected to spread and with the passage of time lead to overland and sea trade routes, including those that brought plant materials from Asia to Europe to meet the demand for spices as seasonings and medicines. New insights into the causal agents of poor health were acquired during the 18th and 19th centuries (such as the germ theory developed by Pasteur and Koch, use of disinfectants by Lister, plus the work of many others), but medicinal and aromatic plants remained the primary pharmaceutical agents into the early 1900s.

Technological innovations and political and social forces at the beginning of the 20th century caused a rapid decline in the use of plants as medicine. The development of sulfa drugs in the 1930s and the synthesis of organic chemicals in the 1940s produced additional sources of medicines and, indeed, became the preferred method for treatment in some countries, especially in America where the 1910 Flexner Report and the Medical Association indicated only trained physicians using allopathic medicines should be allowed to practice medicine. At the same time, Western societies, modernized by the industrial revolution, became infatuated with the social and economic changes following World War I and identified with the new, synthesized chemical medicines, requesting these in place of herbal medicines and resulting in a decline in the use of plants and plant extracts until the mid-1970s. By the middle of the 20th century, medical practitioners and consumers were demanding scientific proof of efficacy through double-blind clinical trials in contrast to the traditional use and mythological associations linked with medicinal plant materials. The decline in the use of herbal preparations as medicines can be inferred from the decline in listing of medicinal species in the U.S. Pharmacopeia and National Formulary beginning in the early 1900s.

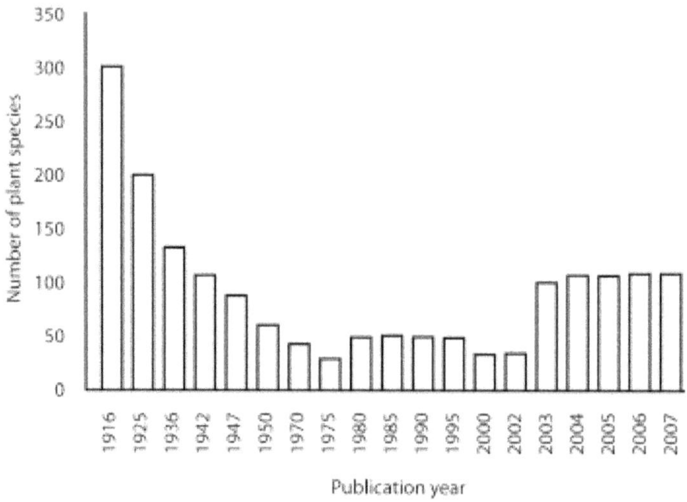

Data collected by a count of plants and plant extracts

At the start of the 20th century, over 40 percent of the listed drugs (mostly crude extracts) originated from plants, but by the mid-1970s, the listing of plant materials had decreased to less than 5% of the drugs in the Pharmacopeia and Formulary. In the intervening years, the American concept of medicine had evolved from a collection of plant materials with a mixture of constituents to a medication containing only one chemical formula. The collection of laws and regulations, originally developed to protect the public from worthless health products, unsanitary manufacturing practices, and unscrupulous sales people, limited access to phytomedicines and medical practitioners that used phytomedicines. With decreased demand, interest in the cultivation of medicinal and aromatic plants in the United States decreased. The US Department of Agriculture, which published cultivation guides for farmers growing medicinal aromatic plants in the first half of the 1900s, failed to provide such guides in the latter half of the 20th century. A sampling of horticultural and garden books published in the early 1900s frequently contains cultivation information on aromatic and medicinal herbs, but similar books published as late as the 1980s mention these crops only briefly and focus on culinary herbs.

The revival of US interest in culinary herbs appears to have been initiated by a revival of interest in natural products in the 1960s and driven by demographic changes in the population. For example, the current population growth rate among those of Asian and Hispanic origin, cultural groups acknowledged to enjoy more highly spiced foods than cultures of European origin, leads all ethnic groups at 61%. In addition, an aging American population began to use medicinal and aromatic plants to stimulate aging taste buds and, for health reasons, as substitutes for salt. Changes to the family structure in which both parents worked led to increased use of prepared foods that contained more spices. As consumers became more worldly oriented and more knowledgeable about health during the 1980s and 1990s, interest in organic and natural foods placed new emphasis on the benefits of using medicinal and aromatic plants. The expansion of this American interest in medicinal and aromatic plants was promoted by observations on the use of alternative medicines in China during President Nixon's visit to that country in 1972.

While conventional medicine establishments generally disapproved of the revival of herbal medicines in the US, consumers began to explore and use these products as evidenced by the enhanced market for dietary supplements during the 1990s, increasing in sales from less than $1 billion annually in the early 1990s to over $4 billion annually by the end of the 1990s. With passage of the Dietary Supplement Health and Education Act, sales of supplements, many of which are herbal products, continued to increase despite warnings about lack of efficacy and health hazards from the conventional medicine system in the US.

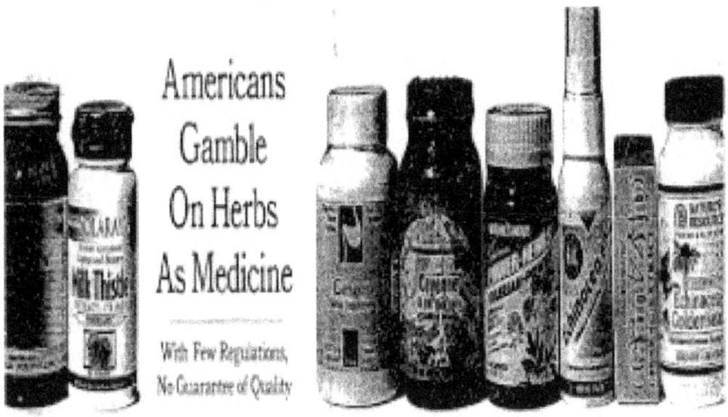

Newspaper warning about use of medicinal plants as medicines

The aging population of the US, which needed increased health care, discovered herbal medicines were an attractive alternative to the comparatively high costs of conventional medicines. The use of herbal medicines in the US by regular and occasional users increased from 2% of the population in 1990 to 37% in 2000. Yet, this development in the US is in contrast to most other countries that had remained committed to the use of herbal medicines as part of their health care system throughout the 20th century. In most areas of Africa and Asia, the traditional healer continues to be the main source of medical care into the 21st century, primarily due to the relatively high cost of conventional medicines and the lack of trained physicians.

Globalization of trade in the late 1990s and early 2000s and concerns about the conservation of genetic diversity affected the cultivation of medicinal plants. As demand for medicinal species in the US grew, domestic and foreign growers increased production. Quality standards for plant material increased with processors and consumers demanding clean (no physical nor chemical contaminating adulterants), consistent, and certifiable (identifiable for origin and history) products. Research on medicinal and aromatic plants initiated in the 1980s and 1990s led to improvements in production of plants, extraction of bioactive constituents, and confirmation of medicinal applications.

2.8.2 Future

Demand for medicinal and aromatic plants in the United States can be expected to continue for the near future, although the rate of sale increases for many medicinal and aromatic plant materials will probably not match those exhibited during the 1990s. While the global market for medicinal and aromatic plants can be estimated to be at least US$60 billion, exact market figures and market trends are difficult to ascertain due to herbal materials in a vast array of products being sold through a large number outlets, ranging from entrepreneurial sales over the internet to mass market sales in supermarkets and natural product stores.

In addition, favorable or unfavorable press reports about a particular herbal product can cause an especially strong growth or a rapid decline in interest and sales. Most market surveys suggest only a slow increase in overall demand within the US for medicinal and aromatic plants, as compared with the 1990s. If the US medical establishment fully accepts medicinal plants as part of the mainstream, conventional medicine system, sales could be expected to significantly increase.

Over the near future, continued globalization of trade and markets along with ethnobotanical exploration can be projected to continue to bring awareness of new plant materials for home, medicinal, and industrial use. In addition to the demand created by population diversification and aging in the past 10 years, the relatively high cost of medical treatment in the US and the failure of conventional medicine to have satisfactory treatments for aliments, such as those associated with obesity and diabetes as shown in table, and currently incurable afflictions, such as HIV, cardiovascular and degenerative diseases, and cancer, have stimulated many healthcare consumers to test alternative medicines.

Table Obesity and diabetes

Designation[z]	Population (%)
Overweight (BMI ≥ 25)	64.5
Obese (BMI ≥ 30)	30.5
Severely obese (BMI ≥ 35)	4.7
Diabetes	7.0

The search for disease cures have led to global cooperation and extended research efforts on the cultivation and improvement of such medicinal plants as Artemisia annual for treatment of malaria, a disease estimated to be responsible for almost three million deaths annually over the past 30 years, as synthesized drugs became ineffective.

Rising incomes in Asia are likely to raise the standard of living of residents, increasing demand for additional medicinal and aromatic plants as the population suffers from the detrimental affects of ageing, weight gain, and other medical problems that frequently occur in relatively prosperous societies. The increase in demand for medicinal and aromatic plants will likely continue to threaten native species in some localities. Price differentials between wild and cultivated plants due to a desire for the wild material or the unavailability of cultivated plant material currently encourages unsustainable collection practices in some localities, especially in economically depressed regions that lack resources for protecting plants.

The financial gains for collecting and selling local plant material frequently represent a substantial share of total income for many medicinal plant collectors in several regions. As an example, collection of wild ginseng in West Virginia can be a considerable addition to budgets of poor families. In many instances, a switch to cultivated species to protect endangered species is problematic due to the difficulty of duplicating the demanding environmental requirements of wild species in cultivated fields and due to the socio-economic impact on native cultures and local economies when cultivation is shifted to large scale production outside the local area.

In addition, consumer concern about protection of endangered species and prosperity of native cultures, factors that may decrease or increase medicinal plant purchases, respectively, will most likely require labeling to demonstrate ethical wild-crafting practices and fair-trade (a guarantee of a just financial return to the grower/ collector for work).Continued loss of habitat in the future due to deforestation and development can be expected to remain a threat to many medicinal and aromatic species in both developing and industrialized countries.

In tropical areas such as Amazonia and West Africa, changing land use from logging, ranching, mining, and agriculture have been identified as responsible for changes in forest composition and structure, frequently creating environments unfavorable to growing native medicinal and aromatic species and posing detrimental affects on traditional healthcare. Such destruction in natural ecosystems and the resultant losses in medicinal and aromatic species will surely increase pressure for preservation and cultivation of endangered flora. Shortages of available plant material for collection in the natural environment of medicinal and aromatic plants can be expected to lead to increased costs for plant material until cultivation systems are in place. Estimates suggest the number of plant species used for medicinal purposes, most of which are collected in the wild, is more than 52,000. In current medicinal and aromatic plant markets, the demand is for organic products, matching the demand for organic foods and suggesting that the current base of customers for medicinal and aromatic plant products are the same as those that purchase organic foods.

As consumers become more involved with health and wellness, future medicinal and aromatic plant markets will need to respond to this demand by consumers for quality plant material, most likely produced sustainably and uncontaminated by either synthetic pesticides or by genetically modified organisms. As processors expand to meet the demand, global trade in medicinal and aromatic plants can be expected to increase to make available certified organic plant material needed for the development of new formulations and marketing concepts. Sales of medicinal and aromatic plant combinations and herbal drinks have posted market gains during the past few years by offering the liveliness and vigor of youth, the promise of vitality, simplicity, and sustainability that can be expected to remain key market concepts in selling medicinal and aromatic products in the near future.

Sales of organic non-food items (most of which could be expected to contain medicinal and aromatic plants or plant extracts) increased by one-third in 2005. As per reports, sales of dietary supplements, many of which contain medicinal plant materials, increased by 4.5% to $21.3 billion in 2005, with sales forecast to grow 1.5 to 2 times faster than that of the US economy. Aromatic plants and plant extracts have become extensively used in flower arrangements and the fragrance industry. The market for candles and home fragrances, many of which contain aromatic plants and/or plant extracts, reached an estimated $8.4 billion in 2004, a growth of 14.1% from 2003 that has primarily been attributed to the popularity of aromatherapy and scented candles. Incense, made from a combination of fragrant gums, resins, woods, and spices has an estimated market value in the US of $17 million ($12.4 million imports and $4.6 million exports).

In addition to a role in traditional and in alternative and complimentary medicinal markets, medicinal and aromatic plants maintain a role in both over-the-counter and prescription drugs in conventional medicine. An estimated 25% of conventional pharmaceuticals are derived from medicinal plants. The estimated global market for plant derived drugs was $18 billion and is expected to grow to $26 billion by 2011, with the US and Canada accounting for over 50% of the market demand. Of new interest is the development of plant-made pharmaceuticals where plants are being used to produce therapeutic proteins that could be used for treating diseases. Currently, common food and feed crops, such as alfalfa, barley, corn, rice, and safflower, are being used to produce the proteins that are subsequently isolated from the plant material. Although these products are not yet on the market, they may offer medicinal plant growers and processors new business opportunities in the future.

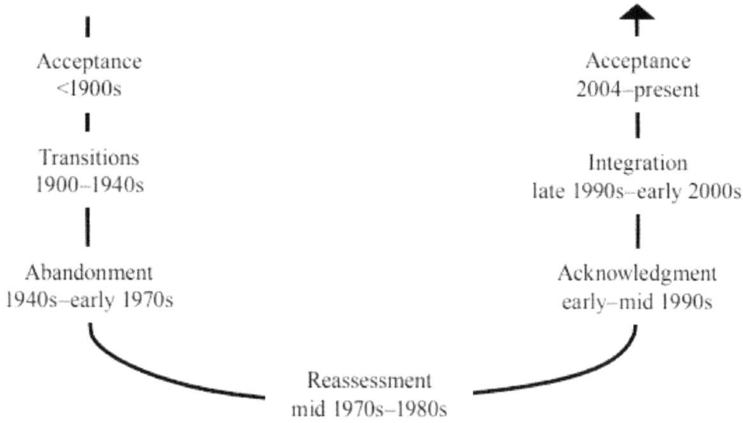

"U-turn" in acceptance of alternative medicines

In the past few years, the conventional American medical system seems to have done a "U-turn" and accepted the use of medicinal and aromatic by patients. Yet, inconsistent product quality, due to genotypic variation within plant species and environmental effects that alter constituent levels and distribution, remains an issue for use of medicinal and aromatic plants that growers and processors can expect to continue to face in the immediate future. Quality is frequently judged by color, aroma, taste, and effect of the plant material, although the levels of various chemical constituents may also be analyzed in facilities equipped or associated with testing laboratories. Additional terminology associated with medicinal and aromatic plants and used in an effort to judge quality aspects include organic (produced according to certification rules, including without the use of synthetic fertilizers or pesticides), wild-crafted (collected in the natural environment with no human contact before harvest), woods-grown (planted in natural environment with protection and care during growth), and commercial (produced with the possible use of synthetic pesticides or fertilizers).

Good agricultural practices, good collection practices, and good manufacturing practices have been developed to help growers, collectors, and processors to produce and maintain quality medicinal and aromatic plant material. In the future, these guidelines are likely to be revised and become one of the standards for quality determination. Such practices should help reduce adulterants and contaminants in medicinal and aromatic plant materials brought to market. Such adulterants (extraneous and fake plant material, counterfeit goods, synthetic drugs, and other non-specific materials) are frequently a health hazard and deprive the consumer of expected benefits.

In world markets, demand for medicinal and aromatic plant materials should continue into the foreseeable future. Current and future changes in demographics (age, culture, incomes, diseases, and other human conditions), public concern about healthcare (availability and expense), and familiarity with plant products can be expected to bring more people to sample and commit to using medicinal and/or aromatic plant products. Acceptance of alternative and complimentary medicines by conventional medical systems should reassure those questioning the use of plant materials, enhance demand for medicinal plants, and help establish a partnership between conventional and alternative medicines for the benefit of the consumer. Rising consumer interest in use of natural and organic products, in protection of endangered species, in intellectual property rights of native populations, and in the value of fair trade will most likely continue, bringing a need to validate plant sources and, in some instances, a preference for cultivation.

Consumer interest in medicinal and aromatic plants is continuing to change in the US marketplace as segments of society become more aware of the possible relationships between good health and healthy living. The concept of Western medicine in which health is defined as absence of disease and all body systems functioning is moderating and becoming more adjusted to the idea of balance within the mind and body. As consumers become better informed about issues of food, health, and nutrition, they also become better informed about the controversies and concerns surrounding conventional medicine, genetically engineered products, pesticide contaminated food, and similar issues. Such consumers frequently choose or move towards a life-style likely to bring them into organic and natural food stores and to try alternative medical care. Increased use of medicinal and aromatic plants will most likely be part of this evolution.

Chapter 3

LANDSCAPING OF ORNAMENTAL CROPS/PLANTS

3.1 INTRODUCTION

Dutch artists of the 16th century used the word "landschap" to describe paintings of scenery on land. The modern English word "landscape" borrows from the Dutch and refers to a visual image of outdoor spaces. Landscaping, therefore, is about visualizing a picture for a space and then creating that space. Landscaping incorporates into the space botanical elements, such as trees, lawns and shrubbery and other flora, along with hardscapes, such as pathways, furniture, planting beds, water features and fences. landscaping involves creating a plan or picture for an aesthetically pleasing outdoor area—with the use of grasses, plants, trees, flowers, and hardscapes, like water features, stones, fences, planting beds, etc.

Like landscaping, gardening involves design, planning and maintenance, but gardening usually involves only the plants in a space. Landscape designers and gardeners can sketch a design for a garden and determine a plant list, but only the gardeners will plant, weed, cultivate, replace and harvest the plants. Gardening is an ongoing activity because gardens require tending and specialized care throughout the year. Gardening also requires knowledge, from experience and education, of several natural processes, including soil ecology, climates, plant physiology, water drainage and pest infestations.

Landscape ecology by definition deals with the ecology of landscapes. Surprisingly, there are many different interpretations of the term "landscape." The disparity in definitions makes it difficult to communicate clearly, and even more difficult to establish consistent management policies. Definitions of landscape invariably include an area of land containing a mosaic of patches or landscape elements.

Forman and Godron defined landscape as a heterogeneous land area composed of a cluster of interacting ecosystems that is repeated in similar form throughout. Turner et al define landscape as an area that is spatially heterogeneous in at least one factor of interest. The landscape concept differs from the traditional ecosystem concept in focusing on groups of ecosystems and the interactions among them – the focus is on spatial heterogeneity and its impact on process. There are many variants of the definition depending on the research or management context.

For example, from a wildlife perspective, we might define landscape as an area of land containing a mosaic of habitat patches, often within which a particular "focal" or "target" habitat patch is embedded. Because habitat patches can only be defined relative to a particular organism's perception and scaling of the environment, landscape size would differ among organisms. However, landscapes generally occupy some spatial scale intermediate between an organism's normal home range and its regional distribution. In-otherwords, because each organism scales the environment differently, there is no absolute size for a landscape. From an organism-centered perspective, the size of a landscape varies depending on what constitutes a mosaic of habitat or resource patches meaningful to that particular organism; a landscape could range in absolute scale from an area smaller than a single forest stand (e.g., a individual log) to an entire ecoregion. If you adopt this organism-centered definition of a landscape, a logical consequence of this is a mandate to manage habitats across the full range of spatial scales; each scale, whether it be the stand or watershed, or some other scale, will likely be important for a subset of species, and each species will likely respond to more than 1 scale.

3.1.1 Concept

Regardless of how landscape is defined, the "concept" of a landscape is unequivocal. All landscapes have a user-defined structure (pattern) that is hypothesized to influence its function (process). This interaction between spatial pattern and process defines the landscape concept. Landscape structure.–The structure of a landscape is defined by the particular spatial pattern being represented, and it consists of two components: composition and configuration. The composition of a landscape is defined by the spatial elements that are distinguished in the map and believed to be relevant to the landscape function under consideration. Composition represents the nonspatial aspect of a landscape, since only number and abundance of landscape elements is considered, not their spatial configuration. The configuration of a landscape is defined by the spatial character, arrangement and context of the elements. Configuration represents the spatial aspect of a landscape. Together these two components define the spatial pattern or heterogeneity of the landscape.

Landscape function.–The function of a landscape is defined by the phenomena under consideration and can be a multitude of different things. In general, the services that landscapes provide to humans are functions and include things like providing for biological diversity, recycling nutrients, sequestering carbon, producing clean water, etc.

Landscape boundary

Landscapes do not exist in isolation. Landscapes are nested within larger landscapes, that are nested within larger landscapes, and so on. In other words, each landscape has a context or regional setting, regardless of scale and how the landscape is defined. The landscape context may constrain processes operating within the landscape.

Landscapes are "open" systems; energy, materials, and organisms move into and out of the landscape. This is especially true in practice, where landscapes are often somewhat arbitrarily delineated. That broad-scale processes act to constrain or influence finer-scale phenomena is one of the key principles of hierarchy theory and 'supply-side' ecology. The importance of the landscape context is dependent on the phenomenon of interest, but typically varies as a function of the "openness" of the landscape. The "openness" of the landscape depends not only on the phenomenon under consideration, but on the basis used for delineating the landscape boundary.

For example, from a geomorphological or hydrological perspective, the watershed forms a natural landscape, and a landscape defined in this manner might be considered relatively "closed". Of course, energy and materials flow out of this landscape and the landscape context influences the input of energy and materials by affecting climate and so forth, but the system is nevertheless relatively closed. Conversely, from the perspective of a bird population, topographic boundaries may have little ecological relevance, and the landscape defined on the basis of watershed boundaries might be considered a relatively "open" system.

3.1.2 Sources of New Native Ornamental Plants

"Xeriscape" has often been interpreted by the general public to mean "zero-scape" and was equated with severe minimalist drought tolerant plantings. The terms "sustainable landscapes" and "appropriate horticulture" better convey the full intent of "xeriscape." Native plant materials are a vital and viable element of the future of xeriscape landscapes and gardens across America and are the primary focus of this paper. As more gardens and landscapes are designed and installed true to their local conditions, opportunities will continue to expand for the use and development of new native plants for ornamental horticulture.

The word xeriscape was coined in 1981, to capture the idea of water conserving landscapes but conceptually xeriscape involves much more. There are seven essential components of xeriscaping: appropriate planning and design; appropriate/minimal turf areas; appropriate soil preparation and analysis; appropriate plant selection; appropriate watering methods; appropriate use of mulches; and appropriate landscape maintenance. In essence, the underlying concept behind xeriscape is the doctrine of regionally appropriate horticulture.

In practice, however, xeriscape often has been reduced to the landscaping idea of maximizing the efficient use of water in gardens and landscapes by creating areas which group plants by their water requirements. Typically this has been interpreted as growing the plants with the greatest water need nearest the home and transitioning to very low to no irrigation at the zone furthest from the home. This zoning of the landscape according to water use may also have additional benefits pertaining to fire breaks, wildlife value, and in enhancing the diversity of plants that may be grown in the garden.

History

Due to the dominance of our primarily temperate European horticultural heritage, there has always been a predisposition toward water-loving exotic plants in our landscapes and gardens. The westward expansion and settling of the country brought these same traditions and plants into increasingly dry and warmer regions that are prone to long periods of drought. As long as water was cheap and abundant there wasn't a problem and a more or less standard set of coast-to-coast plants thrived in spacious green havens. Over much of the country, water is the limiting resource governing the use of ornamental plants in gardens. The huge increase in population in the U.S. Southwest has in many areas outstripped local water supplies. In some areas of the country, water for landscape or garden use is seasonally rationed or restricted. The cost of water, even during years of abundant rainfall, is always increasing. Water quality issues, including salinity and contamination by nitrates and bacteria, lead to increased costs due to necessary water treatment programs. As time passed, our use of gardens and landscapes changed from seasonal retreats and visual frames for our homes and buildings to heavily used outside multi-purpose spaces. These changes resulted in the establishment of the xeriscape concept.

Trends

There are a number of trends in home landscapes and gardens which will influence the ornamental plant industry. Those which considered to be among the most important are:

1. the decrease in the amount of time, money and expertise that most households have to invest in properly planting and maintaining gardens;
2. the shrinking in average size of new gardens;
3. the increasing use of gardens as living spaces and outdoor rooms;
4. the increasing cost of water, labor, fertilizers, and chemicals;
5. the restriction or limiting of water use for garden and landscape purposes.

Although there is always a market for colorful annuals, biennials, and short-lived perennials, these are generally more difficult groups to depend upon in landscapes and gardens. Longer lived perennials, subshrubs, shrubs and trees are the backbone of landscapes and gardens and will constitute the primary nursery market.

Under current conditions, I am convinced that a greater diversity of plants which flourish under existing climatic conditions across the country are needed. These changes should be viewed as potential opportunities and require the search for, and development of, perennials, subshrubs, shrubs and trees in the following categories: dwarf or compact forms of existing plants; new plants of small stature; plants that are all season performers; low maintenance plants; long, or continuous, seasons of color, be it flowers, foliage, fruits, stems or bark; good-looking, water-thrifty plants; and plants that attract wildlife.

Native plants which fall into any or all of these categories will be particularly desirable as they have the additional benefit of fitting into the local and regional conditions both visually and ecologically. This opportunity creates a new, relatively unexploited regional niche market for the nursery industry, and should eventually lead to a more striking regionalism in the American landscape vernacular.

There are a number of people and institutions throughout the country looking at our native plants as a direct or secondary source of new introductions for the nursery industry, but anyone can become involved in this process. The methods employed to create new native ornamentals can be as informal as selection from the wild or as sophisticated as plant breeding including even genetic engineering.

A good place to begin the investigation of new native plants for drier portions of our landscapes is to look at the large plant families and genera in the region you are serving, and to concentrate the search on those plants inhabiting drier habitats. In California and the southwest, these would likely include some of the following: Agavaceae (Agave, Nolina, Yucca), Alliaceae (Allium), Amaryllidaceae (Brodiaea, Dichelostemma, Triteleia), Asclepidaceae (Asclepias), Asteraceae (Artemisia, Erigeron, Senecio, Haplopappus--Ericameria, Hazardia, Isocoma), Cactaceae (Opuntia), Crassulaceae (Dudleya, Sedum), Cyperaceae (Carex), Ericaceae (Arctostaphylos), Fabaceae (Astragalus, Lupinus), Fagaceae (Quercus), Grossulariaceae (Ribes), Iridaceae (Iris, Sisyrinchium), Hydrophyllaceae (Phacelia), Lamiaceae (Lepechinia,

Monardella, Salvia), Liliaceae (Calochortus, Lilium), Malvaceae (Malacothamnus, Sidalcea, Sphaeralcea), Poaceae (Festuca), Polygonaceae (Eriogonum), Ranunculaceae (Aquilegia, Delphinium, Ranunculus), Rhamnaceae (Ceanothus, Rhamnus), Rosaceae (Adenostoma, Cercocarpus, Heteromeles, Holodiscus, Prunus, Purshia, Rosa, Spiraea), Saxifragaceae (Heuchera), and Scrophulariaceae (Penstemon).

The majority of these large genera and families have yet to be tested or evaluated for horticultural purposes in a serious systematic fashion including most perennial members of Phacelia, Monardella, Eriogonum, and Astragalus. A number of these large genera have well known propagation problems (such as commercially viable method for asexual propagation of Quercus and Calochortus) or are known to be especially difficult to grow and/or maintain under nursery conditions (like Calochortus and many native Delphinium and Lupinus species).

The selection of an individual plant for possible introduction is generally accomplished by finding a desirable feature or set of features that are deemed desirable. These individuals can be found in the wild or in the garden and often represent extremes of the natural variation: compact growth habit, different foliage, albinos, and unusual color forms. Additional factors such as mutations including witch's brooms, variegated foliage, and natural hybrids also contribute to the possible pool of plants to choose from. Traditionally these matters of chance have been the most frequently reported sources of many excellent cultivars: Heterotheca (Chrysopsis) villosa 'San Bruno Mountain', a sterile, free-flowering dwarf selection; Artemisia pycnocephala 'David's Choice', a heat tolerant compact selection; Erigeron 'W.R.', a compact, free-flowering, heat and cold tolerant chance wild hybrid between Erigeron glaucus and an unknown species; and Acer macrophylla 'Seattle Sentinel', a fastigiate selection.

Another especially rich area for horticultural selection can be found in nearly all taxonomically confusing groups at either the generic [example: Arctostaphylos and Zauschneria (Epilobium)] or specific [example: Mimulus aurantiacus or Mahonia (Berberis) aquifolium] level. Taxonomic chaos is often indicative of the extreme plasticity found naturally in the group and by inference the potential horticultural availability of variation to select from. Similarly, there are a number of extremely variable species that are nearly as ripe for selection as the taxonomically complicated groups listed above. These species include Rhamnus californica, Erigeron glaucus, Ceanothus maritimus, Quercus chrysolepis, and Juniperus communis.

Peripheral populations of a desirable plant are another source of native plants for landscapes and gardens. Plants from these populations may be more tolerant of heat or cold, higher or lower elevations, drought or wetness, may exhibit resistance to diseases or pests, or may be adaptable to a different soil type or condition. An excellent example of this phenomena is Arctostaphylos uva-ursi 'Point Reyes', a selection from the fog-bound headlands of the Point Reyes peninsula in Marin County that is extremely tolerant of heat and drought in comparison to typical members of the species. The small relictual San Bernardino Mountains population of Populus tremuloides is remarkably well adapted to growing conditions at low elevations in southern California. When plants from these peripheral populations are not themselves horticulturally desirable, they may carry genes which may be useful in a breeding program.

Another particularly rich source of new plants to investigate are the genera that have been dependably used in landscapes and gardens before, and look for a different species with desirable characteristics. Cultivars of the "new" species can be selected directly or the "new" species may be used to create hybrids with the "established" species.

Breeding programs involving controlled crosses and hybridization between species are relatively rarely encountered in the native flora. There are, however, a few exemplary programs which have yielded many fine plants: the pacific coast hybrid irises, the Mimulus hybrids (of the section Diplacus), the Lewisia cotyledon complex, and Lilium hybrids.

All new cultivars should be named, described, published and registered with the proper registration authority. A list of all currently accepted registration authorities for ornamental plants is available from the American Association of Botanical Gardens and Arboreta (AABGA).

Testing and Evaluation

All too frequently, a promising native plant species or cultivar does not live up to its potential. Almost without exception, these plants are victims of poor testing and evaluation.

Establishing a set of desirable traits and characteristics of plants which the public actively wants or desires is a critical first step in the development of a new plant introduction strategy. Consulting with focus groups from the nursery industry, the gardening public, native plant enthusiasts, realtors, horticultural groups and others should provide the researcher with plenty of input on a variety of desirable plant traits. As progress is made in the selection process, the new plants should be planted out in a number of test sites. Test plantings should then be evaluated by a number of outside reviewers. Plants surviving through the testing and evaluation process are then ready to be introduced and marketed to potential user groups.

Conservation of Genetic Resources

Everyone involved with the selection, use and promotion of new native plant species and cultivars and/or any of their close relatives must be fully aware of the potential negative consequences of their use in gardens and landscapes. Most of these concerns center around the issue of conservation of genetic resources and the genetic pollution of native plant populations in the wild. Three examples from the California flora will serve to illuminate this issue.

Nevin's Barberry, an extremely rare plant in nature, is found in widely distributed and ecologically varied small populations in southern California. It is, however, a relatively commonly seen landscape plant throughout the state. A molecular level study of these plants showed that there is little to no variation present in the gene pool. Therefore, the seed source of plants to be planted in the vicinity of the remaining individuals in the wild is not of concern--there will be no adverse consequences to the gene pool.

The California dandelion (Taraxacum californicum) from the wet meadows in the San Bernardino Mountains is a rare plant whose continued existence is threatened by ongoing hybridization with the common European dandelion (Taraxacum officinale), a weedy pest plant that is common throughout the range of the rare species.

The Monterey pine (Pinus radiata) is known from five geographically isolated populations (three are from central coastal California and two are on islands off Baja California). This tree is a significant forest tree in the southern hemisphere, particularly in New Zealand. Material from all five populations has been grown together on a massive scale in New Zealand, such that the primary seed source of Monterey pine (even in California) is from New Zealand. The issue of primary concern here is that the "mongrel" Monterey pines of New Zealand mixed origin will genetically pollute the "pure" native stands of these trees, such that the unique genetic character of the California populations will be lost due to homogenization of the gene pool over time.

3.2 SOIL PREPARATION AND PLANTING PROCEDURES IN LANDSCAPE

3.2.1 Surveying the Planting Site

Before planting, survey the site for potential hazards to plant growth. For instance, new construction sites are often littered with pieces of mortar, plaster or limestone, creating an alkaline soil condition and inhibiting a plant's ability to absorb nutrients. Chemical spills, such as motor oil or gasoline, can also impair plant growth. It may be necessary to remove the top 6 to 8 inches of soil and replace it with a good grade of topsoil. Compacted soils also inhibit root growth.

Ornamentals can be grown on poorly-drained soils if they are planted on raised beds.

3.2.2 Selecting Plant

Always purchase fresh, high-quality plants. Poor-quality plants are not a wise investment. Most ornamental trees and shrubs marketed today are grown and sold in containers, although field-grown plants, sold balled-and-burlapped or packaged bare-root, are also available during certain times of the year. Container-grown plants should have healthy, vigorous tops and white feeder roots on the outer edge of the root ball. Do not be timid about inverting a few plants, removing their pots and examining their roots. Container-grown plants generally transplant well throughout most of the year with minimum shock, although fall and winter months are the best time to transplant.

Woody ornamentals for the landscape are commonly sold three ways: container-grown (left), balled-and-burlapped (center) and bare-rooted (right).

Large trees and shrubs grown in the field are often sold balled-and-burlapped. Because a large portion of the root system is destroyed during digging, they transplant best during the cooler months (October through April). Some trees are grown and marketed in fabric bags, and can be transplanted throughout the year, although the fall and winter months are best.

Packaged bare-root plants, such as roses, should have plump, healthy stems and good root systems that are kept moist in a packing substance like sphagnum moss or wood shavings. The best planting time for these plants is from December to mid-March.

3.2.3 Holding Plants Until They Are Planted

If plants cannot be planted right away, place them in a shaded area and keep the roots moist. If balled-and-burlapped or bare-root plants must be held several days before planting, cover their roots with sawdust, pine straw or soil to conserve moisture. Avoid placing the roots in water or buckets for long periods of time because they will suffocate. Container plants may need daily watering. Make sure plants are well watered before planting and ensure the root ball is thoroughly wet. A dry root ball is difficult to rewet after transplanting.

3.2.4 Planting in Individual Holes

The old adage "never put a ten-dollar tree in a two-dollar hole" applies when planting individual trees and shrubs. Research at the University of Georgia has shown that a large planting hole – at least twice as wide as the root ball – encourages rapid root growth and plant establishment. Dig the planting hole only as deep as the root ball. If the hole is dug deeper, backfill it with soil as necessary and tamp it firmly to prevent settling. Make certain the top of the root ball is level with the soil surface. Some landscape professionals plant the top of the root ball 1 to 2 inches above grade if they know the soil is likely to settle slightly.

Dig the planting hole two times wider than the root ball. Make certain the top of the root ball is level with the soil surface.

Research has also shown that it is not necessary to add organic amendments, such as peat moss, compost or leaf mold, to the planting hole. Organic matter can act like a sponge in the planting hole, absorbing and holding too much moisture and causing the roots to stay too wet. When planting just one plant, it is best to backfill with the same soil removed from the hole. Be sure to break apart any clods and remove stones or other debris before refilling the hole.

Before planting balled-and-burlapped plants, cut any wire or cord from around the trunk and pull back the burlap from the top one-third of the root ball. This will allow newly formed feeder roots to grow into the new environment. When planting on poorly drained soils, remove the burlap completely. When planting trees or shrubs grown in fabric bags, remove the entire bag before planting.

To eliminate air pockets, water the planting site as the backfill soil is placed in the hole. Use your hand, not your foot, to gently firm the soil around the roots. Water thoroughly when finished and water again several hours later.

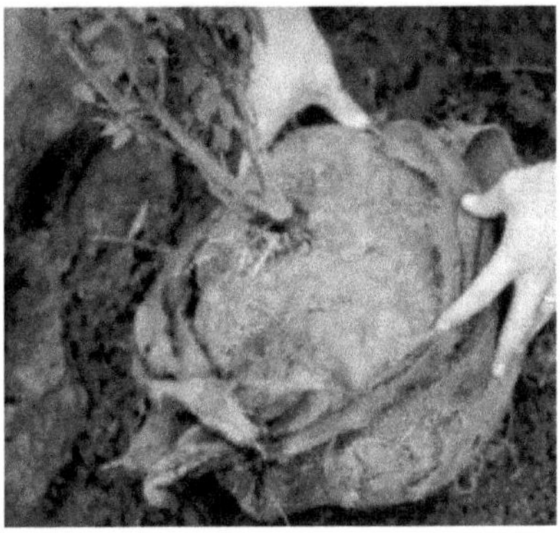

This balled-and-burlapped plant, with the cord cut from around the trunk and the burlap pulled back, is ready for planting.

Slow-release or liquid fertilizers can be added to the planting hole, but granular general-purpose fertilizers, such as 8-8-8 or 10-10-10, can damage tender roots. Wait until the plants are established before applying a granular general-purpose fertilizer.

Finally, shape a small ring of soil, 2 to 3 inches high, along the perimeter of the planting hole. This forms a saucer on top of the soil, which directs water to the roots and prevents runoff.

Finally, uniformly apply a 3-inch layer of mulch over the soil surface. Mulches promote rapid rooting by maintaining uniform moisture levels and temperatures in the soil and by preventing weed competition. Landscape fabrics can be placed under the mulch to help prevent weeds and to conserve moisture.

3.2.5 Planting in Beds

A group of ornamental plants in one area of the landscape will grow more uniformly when planted in a well-prepared bed rather than in individual holes. Begin by deep tilling to a depth of 12 to 15 inches. Then incorporate about 1 pound (2 cups) of an eight to 10 percent nitrogen fertilizer, such as 8-8-8 or 10-10-10, over every 100 square feet of bed area. Only incorporate lime into the bed if the soil test recommends it. After preparing the soil, follow the planting procedure recommended for planting in individual holes.

When planting a group of ornamental plants in the landscape, prepare a good bed by deep tilling to a depth of 12 to 15 inches.

3.2.6 Planting Annuals and Herbaceous Perennials

To achieve the best color displays, annuals and herbaceous perennials must have good drainage, adequate nutrients and available water at all times. Begin by deep tilling the native soil to improve its structure and to ensure good drainage. Then, elevate the bed 6 to 12 inches by adding soil amendments. A raised bed not only ensures good drainage, but also improves the visibility of the color display.

Plant annuals and perennials on raised beds to ensure good drainage and improved visibility.

The type and quantity of soil amendments used depends on the structure and texture of the existing soil, and whether amendments have been previously added to the site. A combination of composted organic matter, composted animal manure and large-particle sand, such as Lithonia granite, are frequently used to amend beds. If bagged organic amendments are used, apply one 40-pound bag per 100 square feet of bed area and incorporate it to a 6- to 8-inch depth. An ideal soil is moist, yet well drained.

Containers let you put splashes of color where you want it, regardless of soil type, and offer an alternative to flowerbeds.

Slow-release fertilizers, such as Osmocote, are excellent for flowerbeds because they give the plants an even supply of nutrients throughout the growing season. Several formulas are available, although one with at least an eight- to nine-month release duration is recommended. Follow manufacturer recommendations for application rate.

After planting, apply about 3 inches of mulch on the soil surface to conserve moisture and prevent weeds. Fine-textured mulches, such as pine straw or pine bark mini-nuggets, stay seated better on the bed than coarse-textured mulches. Finally, water thoroughly. A liquid fertilizer can be applied with the water at planting to provide some immediate nutrients.

3.2.7 Staking and Guying Tree

Protective staking may be necessary for young trees, trees less than 4 feet tall, or trees planted in high foot-traffic areas, such as school grounds or shopping centers. Stakes also protect young tree trunks from lawn mowers and weed eaters, which can severely damage the bark. An area of mulch around the tree is an alternative to protective staking.

Trees with a trunk diameter greater than 1 inch and a height exceeding 4 feet need staking or guy wires to hold them in place until they are established. Staking or guying a tree keeps it from blowing over and uprooting during establishment. Trees with a trunk diameter up to 3 inches can be supported by two to four stakes, depending on the size of the canopy. After they are placed in the ground, the height of the stakes should equal the height of the lowest scaffold branches. Place the stakes along the perimeter of the planting hole, and pound them several inches into the ground to hold them firmly in place.

Secure the tree to the stakes with strong, 12-gauge wire. Attach the wire just above the lowest scaffold branches. Place the wire encircling the tree in a piece of old garden hose to prevent bark injury. Use three guy wires for trees larger than 4 inches in trunk diameter.

Give the tree some slack so it can move slightly with the breeze. Research indicates that a tree allowed some movement during establishment develops a larger root system and stronger trunk than one that is kept stationary.

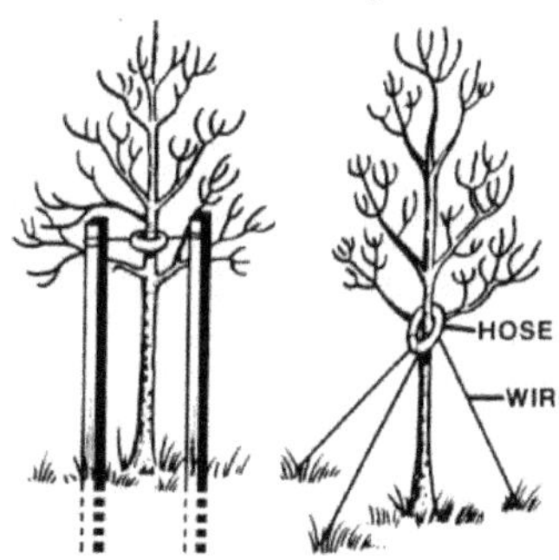

Trees with a trunk diameter greater than 1 inch and a height exceeding 4 feet usually require staking or guying.

Remove stakes and guy wires four to six months after planting to prevent girdling and trunk injury. Remove stakes from fall-planted trees at the start of the spring growing season and from spring-planted trees at the end of the summer growing season.

An alternative staking method that is becoming more popular with arborists and landscape professionals involves placing four wooden stakes (2 x 2 inches wide and 4 feet long) opposite one another just outside the root ball, leaving 4 inches of each stake protruding from the soil. Screws are used to secure two additional 2 x 2-inch stakes to the vertical stakes. This technique prevents the root ball from rocking during establishment, and does not harm the bark of the tree like guy wires sometimes do.

3.2.8 Trunk Wrapping

Trees often have their trunks wrapped to prevent injury during transport from the nursery to the garden center. When purchasing a tree with its trunk wrapped, remove the wrapping at planting time or shortly thereafter. It is not necessary to wrap the trunk of newly-planted trees.

3.2.9 Care of Newly Planted Ornamentals

Watering: Regular watering is critical during establishment of newly planted trees and shrubs. Keep the root system moist, but not too wet, for the first six to eight weeks after planting. The amount of water and frequency of application depend on the soil type and the type of plant. Trees and shrubs may require watering twice a week when there is no rain. Annuals and ground covers may need daily watering during establishment.

Fertilization: There are many slow-release fertilizers on the market that feed plants from six to 12 months with one application. Slow-release fertilizers generally cost more than general-purpose fertilizers, but they require fewer applications. Follow application guidelines on the bag or container.

If you use general-purpose fertilizers, use light applications for newly-planted ornamentals during the first growing season. For shrubs less than 12 inches tall, apply 1 level teaspoon of a 12 to 16 percent nitrogen source (12-4-8 or 16-4-8) or 1 level tablespoon of an eight to 10 percent nitrogen source (8-8-8 or 10-10-10) three times during the growing season (March through September). Broadcast the fertilizer along the perimeter of the planting hole.

Give trees 2 tablespoons of a 12 to 16 percent nitrogen source for each inch of trunk diameter three times during their first growing season. Broadcast fertilizers evenly over an area extending 6 inches from the trunk to 1 foot beyond the branch spread or canopy.

Newly planted ground covers benefit from a complete, balanced fertilizer, like 8-8-8 or 10-10-10. An application rate of 1 pound per 100 square feet is sufficient. When broadcasting fertilizers over the top of the foliage, be sure the foliage is dry, and water soon after application.

3.3 PRINCIPLES OF LANDSCAPING

Landscaping is a unique art that combines a sense of aesthetics and creativity to form visually pleasing scenery. One of the most important pr-requisites of a landscaper is creativity; without creativity there would be no originality to the designs. A creative person can set about with designing a garden following some simple rules. You must keep in mind some important principles before beginning with the task at hand. A garden can be designed using original ideas or on "borrowed ideas". The basic principles of landscape designing give a general guideline to the art of designing a garden. This doesn't imply that every principle should be applied to every part of a garden plan. An effectively understanding of these principles will actually help in aiding your creativity. It is said 'beauty lies in the eyes of the beholder. A garden has immense potential to turn into a work of art thing. These principles will help define the intricate elements of beauty, you must know a bit about them even if you only wish to make a few minor changes to please your creative whimsies.

3.3.1 Unity

Unity is regarded as the most important goal while designing a garden. Unity refers to a sense of consistency and repetition to the entire lay out of the garden. Repetition is a vital element that contributes to the principle of unity. Similar elements like plants, plant groups, or decor are repeated at regular intervals in order to give a sense of symmetry to the garden. Consistency creates unity since it creates the impression that some or all the elements of the garden blend in together to make one whole. Unity can be attained by ensuring appropriate consistency of character of the various elements in the designing of the garden. By character, a reference to the texture, height, size, schemes, color, etc. of different elements is indicated. A good illustration to highlight the importance of unity would be the use of accent boulders at regular intervals and not erratically. The boulders should preferably be made out of the same rock.

This principle also applies to all the other elements such as groups of plants and other accessories that may add beauty to the garden. An easy method of creating unity in your garden is by focusing the design around some common theme. One can make adequate use of garden décor since it is one of the most important elements of a theme. For example, if you love butterflies, you may wish to make use of a theme of plants that attract butterflies as well as other delicate décor items like statues, ornaments related to butterflies. Unity should be effectively articulated through at the very least one element in your landscape and preferably many more. Unity symbolizes harmony and if your design radiates a feeling of harmony, it will create a sense of peace and tranquility.

3.3.2 Simplicity

This is one of the mot elementary principles of art and design. It's one of the best guidelines that could be regarded as a principle that could be kept in mind by a beginner. A simple design is easy to achieve and you will be able to do it yourself without much efforts or help from external sources. A complicated design is the work of an expert; if you venture into trying anything difficult, you may end up not achieving your real goals. Simplicity in choosing the right plants is probably the best option you have; for example, select at the most two or three colors and use them repeatedly throughout the garden or landscape. Keeping the decor to a minimum level and within a per-decided specific theme is one way of bringing about simplicity and elegance in any design.

3.3.3 Balance

Your design should be 'balanced' just as the word implies, it must appeal to your sense of equilibrium. Balance brings about a sense of equality. There are essentially two types of balances in landscape design; the symmetrical balance and the asymmetrical balance.

Symmetrical balance is a definite layering of equally spaced coordinated elements of the garden design. A garden is thought to be equally divided, and thus, both the sides of the garden can share all or a part of the same shape, form, plant group, plant height, bed shapes, colors, theme, etc. You may even rekindle old memories of school times as a child learning art or craft. If you splash color on paper and fold it into half, you ill observe an interesting pattern upon unfolding the paper. Symmetrical design is somewhat like that, it is actually mirror image of a design. So, if you are opting for a symmetrical style of gardening, keep in mind that this would require the designer to 'divide' the garden into two, and design each half similarly.

Asymmetrical balance is the exact opposite of symmetrical balance. It is regarded as one of the most difficult styles of landscape design. The principles of this style of landscape design include colors, forms, textures, etc. used in unique combinations to create just a hint of unity. However, most elements of the garden are used randomly. Asymmetrical design may be better considered as actually being abstract, unbalanced, or free form while still maintaining unity through some form of repetition.

A good example of asymmetrical design is a layout where bed shapes are kept diverse on either side of the line that divides the landscape while preserving some common features like the presence of similar elements and plants. One side can be curved to impart a 'sense of flow' while the other side of he garden can be kept straight and completely opposite to that of the first one. Unity and balance can be achieved through other elements like the color scheme or by proper garden decor.

3.3.4 Contrast

Contrast can be a very interesting feature that defines a garden and lends a pleasing appearance to the entire lay out. Contrast is actually achieved by means of asymmetrical patterning of the garden. For example, flowing lines can be appealing to the eye but straight lines combined with form a very bold contrast. One excellent way to achieve such a startling contrast is by arranging the garden asymmetrically. Asymmetrical arrangement does not depend on the shape of your garden though it can be made to be dependent on the size of the garden.

A good example of striking great contrast would be garden where one side is mostly lined with huge shady trees while the other side has a low growing flower garden. Landscaping can be opined an abstract art form that manages to maintain a sense of harmony and unity. What is really important is that the intensity of contrast remains the same; it could either be soft contrast or bold contrast.

Contrast and harmony can also be achieved using different varieties of plants as well. A good example of contrast is making use of fine foliage instead of foliage that is coarse, rounded leaves in place of spiked. A good idea would be to use complementing or contrasting colors. For example, a flower bed having alternating rows of yellow and red flowers is a bold contrast while a line of red lowers followed by one of pink ones signifies soft contrast. Plant color, texture and height can be blended in one area as against the other area, but the lay out of each area should stay unswerving within the parameters of its own theme. In fact, many beginners prefer to choose a theme to lend and identity the garden. For example, a garden can designed to resemble a lovely and colorful butterfly park.

3.3.5 Color

Color is the most important of all the principles of landscaping. It is color that defines a garden. It is a well known fact that colors are capable of affecting a person's moods. So, a cool color scheme will give a serene appearance to your garden. Color patterns are known to add the aspect of creative reality and interest to the landscape. Bright colors such as shades of orange, yellow and red give an impression of advancements towards you and can make you feel as if an object seems is closer than it really is. Colors such as violets, blues and greens give the impression of moving away from you and make objects seem farther than they really are. Colors like grays, blacks, and whites are thought of as neutral colors and are best used in the background to complement the bright colors in the foreground. However, if you wish to impart a feeling of depth to your landscape coarse textured and dark colored plants in the forefront and fine textured, light plants in the backdrop serve as perfect solution for this.

Colors can also be used to draw attention to specific areas of the garden and at the same time, draw our attention away from undesirable details. A bright hue of colors amongst cooler colors will act as an eye catcher. It is imperative to note that some colors may not blend well together, it is best to avoid using such colors in the same place. For example, yellow and pink may not look good together, so avoid flower beds of such typical colors.

3.3.6 Natural Transition

Natural transition is a term coined together using the two words 'natural' and 'transition' each depicting a vital concept behind the art of designing a garden. It indicates a shift from one style or element of gardening to another one in a smooth manner. It is a technique of making ' changes' to the garden without bringing about rapid and abrupt deviation from the central theme. Transition is basically introduced gradually in order to maintain the touch of discipline to the entire garden décor. It can best be demonstrated in terms of plant color or height, but can also be extended to all elements in the landscape design that can not be limited in any sense owing to external attributes like textures, size, shape or color. In short, transition is a layering technique that can be realized by the arrangement of different element each having unique attributes in a steady, descending or ascending manner. A good example of an effective transition would be a staircase effect where a garden is patterned by arranging large trees next to medium trees which are followed by shrubs which in turn are followed by bedding plants.

Illusions in the entire lay out of the landscape. For example, a transition from taller plants to shorter plants creates a sense of depth and distances making the garden seem bigger than it actually is. A transition from shorter plants to taller plants is used to impart a feeling of nearness by attempting to create a focal point that makes the garden look smaller than it actually is.

3.3.7 Line

Line refers to one of the structural principles of landscape design. It is of great help in determining how the entire garden should be patterned and aligned. Line refers to a sense of flow that a construction depicts; it is a creative architect's aid for designing beautiful gardens. It can mostly be associated with the way beds, pathways, and entryways move and continue. This flow of design is essential in lending an 'aesthetic feel' to your garden. Straight lines are forceful and can be effectively used to represent some strong theme. Curvy lines have a flowing effect and can be used to design a garden that offers a sense of peace and tranquility.

3.3.8 Proportion

Proportion refers to the size of the elements in relation to each other. Of all the principles of landscape design, this is simple enough to understand but is often overlooked. Proportionate garden designs require elaborate thinking and planning. Many elements of landscape design can be deliberately aligned to meet what is actually defined as right proportion.

As an illustration, if you are planning to create a small courtyard garden, a massive garden statue located in the center would make the design seem out of proportion. A small four foot cascade and pond positioned at the center of a large garden would lose its charm since it would be hardly visible. Proportion is a relative concept and elements of a garden can be scaled to fit by creating different rooms in the open expanse. The goal is to create visually pleasing scenery that used various elements of gardening appropriately.

A water expanse if planned as a part of the garden should be structured on the basis of the total area of the garden. For example, a back yard garden that is not too large and has a small pond in the middle. You can arrange various plants around these water bodies. Also, special thought should be given to a proper selection of plants. It is best to avoid using plants that may be out of proportion on account of their color, size or shape.

3.3.9 Repetition

Repetition is a feature that is directly allied to unity. This principle indicates that one typical element of garden design can be used over and over again in a particular area of the garden. It is always good to have an assortment of elements and forms in a garden, but repeating these elements gives a sense of variety to expression. Unity in the garden design is achieved by repeating objects or elements that may be alike. Too many disparate and dissimilar objects can make the garden look disorderly and unplanned. However, this repetition should be controlled to a certain extent. It is possible that too much of one element will make the garden look uninteresting. An ideal solution would be to repeat two or more formations in your garden to make it look pleasing to the eye.

3.4 LANDSCAPE USES OF TREES, SHRUBS AND CLIMBERS

Planting on a slope presents a particular set of challenges. Poor soil composition, uneven amounts of moisture at the top and base of a hill and runoff are the dominant issues that make hillsides tricky to plan.

For any gentle slope that is 3 to 10 percent, select plants that will provide some erosion control. On slopes greater than 10 percent, you may need to incorporate retaining walls, terracing, boulders and other landscape tools such as erosion control blankets and turf reinforcement mats. There's also a safety issue: Slopes greater than 20 percent are too dangerous to mow (or even to weed if the surface is wet and slippery), so plantings should be low-maintenance.

The type of plants you choose are equally important. Annuals, such as lantana, add color but do little for keeping the soil in place in the long run. Many perennials, such as delosperma, provide some erosion control but can be slow to establish. Shrubs and trees can be used, but you must select the right types. For example, if you install an evergreen with a tap root, they absorb a lot of moisture but don't hold the soil in place very well; soil erodes over time and the plant eventually dies.

3.4.1 Shrubs

There are numerous types of landscaping shrubs. They can range in size from the smaller forms on up to the larger tree-like varieties. There are evergreen shrubs, which retain their color and leaves year round, and there are deciduous shrubs, which eventually lose their foliage after putting on a colorful fall show. Read on to learn more about landscaping shrubs and their uses.

Shrubs have many uses within the landscape such as foundation plantings, privacy hedges, and specimen plants. They can be used as backdrops, to define garden beds, and to simply add seasonal interest. Foliage differs from plant to plant, so look for foliage colors that will create a pleasing visual effect and subtle contrast to the other plants within the landscape. Fall foliage colors can range from silver, gold, and variegated varieties to red, purple, and dark green. The best time to plant landscaping shrubs is in the fall or early spring, if necessary. This gives roots plenty of time to become established within the landscape, upping their chances of survival. Before placing any shrub in the landscape, you should first consider the overall height and width of the shrub once it reaches maturity. There should be adequate space available for the shrub to grow. For instance, if a shrub is being used as a foundation planting, ensure that its mature growth will not block windows or obstruct walkways. Also, consider the appearance of the shrub after its initial blooming period has ceased. Since the majority of shrubs are non-flowering most of the year, try to match them to the conditions of your particular landscape. It is also very important that landscaping shrubs complement their surroundings. Therefore, it is also a good idea to research native shrubs as well as landscaping shrubs in order to become familiar with their growing conditions, especially in your given area.

Common Landscape Shrubs

Some of the most common landscaping shrubs include:

- ☆ Azaleas/Rhododendrons – Azaleas and rhododendrons come in a multitude of colors and their striking blooms are a welcome sight in spring. While they typically have a short-lived blooming cycle, many evergreen varieties will provide year-round interest. These shrubs are good choices for foundation plantings, borders, shade and woodland gardens, and specimen plants.
- ☆ Forsythia shrubs – The yellow, spring flowers of forsythia provide much needed color after the dull, darkness of winter. Other than its obvious aesthetic purpose, this easy-care shrub makes a great choice for screening and can even soften corners around the landscape.
- ☆ Hydrangea shrubs – Hydrangeas are also quite popular in the landscape. There are various forms available (pink, blue, and white), and their clustered, summer-flowering blooms are spectacular. Hydrangea shrubs work well in woodland gardens, informal borders, and as specimen plantings.
- ☆ Rose of Sharon – Available in a variety of colors including pink, purple, and white, this shrub not only makes an excellent accent in the landscape but also provides good screening when grouped together along property lines. This shrub can also be used to create lovely backdrops for smaller shrubs and garden plants. Rose of Sharon shrubs may require seasonal pruning to prevent them from having an untamed appearance, especially when planted near the foundations of homes.
- ☆ Boxwood shrubs – Boxwoods are one of the most common evergreen landscaping shrubs. These are oftentimes used as foundation plantings as well as for edging. They are also ideal for creating walled gardens or privacy hedges.
- ☆ Juniper shrubs – Another good evergreen shrub is juniper, with the low-growing variety most commonly used within landscapes. These shrubs are well suited for filling in empty spaces between other, larger shrubs and trees.
- ☆ Holly shrubs – There are many types of holly for the landscape. English holly, which is most common, is an evergreen shrub popular for its spiny-edged leaves and clusters of red berries, especially during winter. While this shrub can definitely stand on its own, it is commonly used in the landscape as a screening hedge or foundation planting.
- ☆ Barberry – The barberry shrub is also commonly used in landscaping. This shrub has thick, green leaves that turn red in the fall and produces

numerous berries as well. It is often used in mixed borders with taller shrubs, perennials, and grasses. Some landscaping shrubs are used strictly for their pleasant aromas.

- These include:
- Lilac shrubs – Lilacs are well known for their fragrant, lilac-colored blooms but they are also available in white and yellow varieties as well. These shrubs have compact forms and those that can get quite tall. Lilacs do well when massed together for screening or as backgrounds for borders. These shrubs can also stand on their own as specimen plantings.
- Witch hazel – While witch hazel looks more like a tree than a shrub, it is definitely meant to be seen as well as smelled. The yellow, fall blooms enable it to blend in nicely with a background of evergreens and should be placed somewhere nearby so its sweet scent can be enjoyed.
- Viburnum shrubs – The intense, sweet clove scent of viburnum shrubs shrubs is often used in foundation plantings or mixed borders.
- Roses – Nothing can quite compare to the aromatic scents given off by roses. While these shrubs are often used simply for aesthetic purposes, they can also be used as attractive privacy hedges and foundation plantings.
- Weigela – Weigela shrubs shrubs bloom profusely throughout spring and fall. Many varieties are available to choose from, including those that are variegated. Besides their stunning appearance, these shrubs can also be used in informal gardens or mass planted for hedges. There are other ways to use shrubs in the landscape. Nearly any shrub can be successfully grown in containers; therefore, even the smallest of landscapes can be home to them, including balconies. They make excellent focal points and are ideal for screening.

3.4.2 Trees

A tree is a tall plant with woody tissue. Trees gather light for photosynthesis through their leaves; this process creates "food" for the tree.

Most of a tree trunk is dead tissue and serves only to support the weight of the tree crown. The outside layers of the tree trunk are the only living portion. The cambium produces new wood and new bark. The band of tissue outside of the cambium is the phloem. Phloem transports new materials (the sugars created from photosynthesis) from the crown to the roots. Dead phloem tissue becomes the bark of a tree. The band of tissue just inside of the cambium is the xylem, which transports water from the roots to the crown. Dead xylem tissue forms the heartwood, or the wood we use for many different purposes. Every year, trees grow two annual rings. In the spring, usually a wider and thinner-walled layer called springwood forms. In the summer, a thicker-walled layer, called summerwood, develops. Annual rings are typical in temperate forest trees.

The first inclination of many homeowners is to plan a formal planting of one row of all the same kind of tree in a straight line, evenly spaced. This ideal is often hard to accomplish. Most plantings attempted this way wind up with one or more trees missing. Sometimes the homeowner tries to replace the missing tree, but the match is rarely perfect. Tree rows are a long-lasting addition to a landscape, so the not-so-perfect effect can be a constant source of frustration for a homeowner who wants the landscape to look tidy. Instead of planting a formal row of all one type of tree, consider planting a mix of different trees and shrubs. By choosing a mixture of plants, you will protect your screen from major loss caused by an outbreak of a single pest or disease. A mixed planting also increases the biodiversity in your landscape by creating habitat for beneficial insects, birds, and animals. By attracting these beneficial creatures, you may find that they successfully keep populations of pest insects in check.

As mentioned above, a formal, straight line planting is difficult to accomplish and often will not meet your goal of screening as rapidly as you would prefer. Consider alternatives to formal planting rows, including the more natural-looking cluster planting method, or multiple staggered row plantings. Cluster planting is done by strategically installing plants in groups of threes, fives, or higher odd numbers to block specific views or prevailing winds. Cluster planting provides an attractive, natural-looking screen without walling off your house and yard like a fortress. By planting clusters away from your house, you also provide backgrounds for interesting flowering and fruiting shrubs that are visible from your deck or living room. Additional cluster plantings can be used to create groves. The combined effect provides screening and an interesting design, allows for good air flow, and accommodates walkways through your property.

Another option is to install multiple staggered rows of trees and shrubs. This type of planting will provide a fortress wall effect. Multiple staggered rows are also commonly recommended for homesteads in open areas to provide protection from prevailing winter winds or reduce noise from traffic or equipment. Multiple staggered rows can be installed using two or three rows of trees and shrubs. The spacing should be based on the average diameter of the plants at maturity. Each successive row is offset or staggered so that the plant fills the visible gap between each of the plants in the previous row. To further enhance the screening, a third row of similar size or smaller plants can be planted. The major benefit of this planting arrangement is that gaps fill in quickly and total screening is achieved more rapidly than a single row. However, a common mistake by the impatient property owner is to plant the trees too close together to achieve immediate privacy. This decision will quickly lead to crowding between neighboring trees, resulting in shading, loss of lower branches, and overall loss of the screening effect.

No matter how you choose to arrange the plants in your screen, keep in mind aesthetics as well as utility. Dot the mini-forest you have created with an occasional focal-point specimen tree--something with unusually colored foliage, unusual fruit, or good fall color. This "punctuation mark" tree will probably cost more than other trees in the screen, but will give you many years of visual enjoyment. Before choosing any plants, evaluate your site and your planting areas. Consider space limitations of the site including utility lines, walkways, driveways, and roads that must remain visible and accessible. Consider zoning requirements for the distance of the plantings to roads and sidewalks. Consider the soil type, drainage, and pH. Also, consider the availability of water, which will be needed until the plants are established or during drought.

More new trees die in landscaping from planting errors than any other cause. All trees should be planted with the root flare just level with the surface of the soil. Use two to three inches of organic mulch in a wide circle, and be sure to keep all soil and mulch away from the trunk. Before you buy any plant, learn more about it (e.g.; major pest problems in the area, cultural requirements, etc.) by asking questions at the garden center or doing research at your local library. Property size and space will also impact your plant choices, so the table is divided into plants suitable for large scale and small scale screens. Finally, deciduous plants are included for situations where summer screening for privacy is the main goal. The final, but perhaps most important step in establishing an attractive screening is proper planting. More new trees die in landscaping from planting errors than any other cause. All trees should be planted with the root flare just level with the surface of the soil. Use two to three inches of organic mulch in a wide circle, and be sure to keep all soil and mulch away from the trunk.

3.4.3 Climbers

Climbers are versatile plants that can often serve the same purpose as shrubs, while using less space. If you have a small garden, climbers can be used to hide walls, or fence off boundaries, without spreading into the limited space of the garden. Climbers can also be used to provide shelter or shade by enclosing structures, such as shadehouses, pergolas and arbours. They are also a cheap and easy way to cover up something unsightly, such as a shed or water tank, or to soften a harsh wall, fence or tree trunk. Many have very attractive flowers or foliage. The only major problems are that, if left uncontrolled, some rampant climbers can block drain pipes or gutters on buildings, while some can do serious damage to the structure (or tree) they are clinging to. Some climbers cause less damage than others, but no matter what you use, you should check for damage every few years, and trim back the climber if need be. If you are worried about climbers causing damage, ones to avoid are those such as Hedera (ivy) and Ficus pumila which cling to even smooth surfaces such as a concrete wall, and can eventually grow into small cracks in the wall then expand enlarging those cracks.

Most climbers need something to climb on. You can use materials such as trellis, wire mesh, single wires or nylon strings. These can be readily attached to:

- A climber on a wall will create a softer appearance and help reduce temperature extremes (eg: keeping the garden cooler on a hot day).
- The supports or roof of a pergola, arbour or archway.
- Between two posts.
- A frame made of galvanised water pipe, or similar material.
- Climbers can also be allowed to climb other plants, or even to scramble over rocks, tree stumps or logs.

Climbers like ivy and Virginia creeper have aerial roots which help them cling to surfaces like walls and fences, but other climbers like honeysuckle and clematis need a framework to scramble over. Wall shrubs don't climb naturally and need training and tying in to a support.

Use a series of horizontal or vertical wires, or a trellis attached to the wall to provide support for your climbing plants. Tie in new shoots with soft garden twine. Garden structures for climbing plants need to be strong and secure as plants can become very heavy with age.

3.5 FACTORS RESPONSIBLE FOR PLANTING OF ORNAMENTAL PLANTS

3.5.1 Introduction

Ornamental plantings on a property help shape and enhance the outdoor environment and experience. Successful, healthy, and long-lived plantings depend on three main factors:

- Proper site analysis and selection of hardy and healthy plants
- Proper site preparation and planting techniques
- After-planting care and maintenance

Underestimating the importance of these three factors can result in disappointment, not to mention expensive, unsightly plantings that grow slowly, die quickly, and are prone to pest infestations. Failing to follow certain fundamental practices is the most common cause of unsuccessful ornamental plantings.

The principles of successful planting do not differ greatly for bare-root, balled and burlapped, and container-grown plants. Each step in the planting process—digging the correct shape and sized planting hole, preparing the planting soil, setting the plant in the hole, watering, and mulching—requires careful attention.

3.5.2 Purchasing Plants

Before you purchase any plants, there are several points to consider.

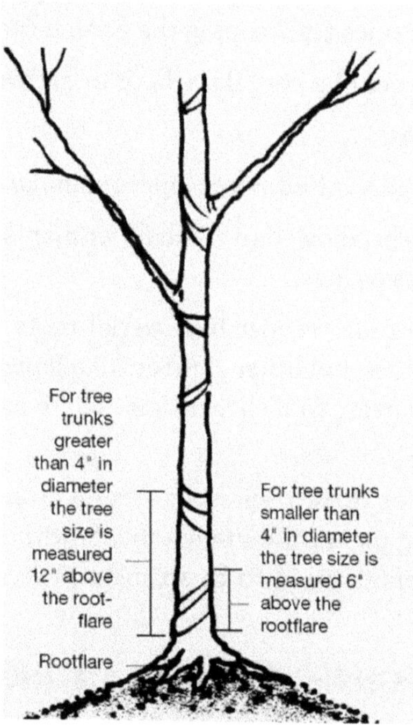

Selecting quality plants at the nursery/garden center.

- ☆ Bare-root material should not have dried out at any time after leaving the production nursery. The entire plant should be alive and fresh-looking with supple, white roots and flexible stems.
- ☆ There should be no large and stubby roots on either bare-root or balled and burlapped material. Fine, fibrous finger-size or smaller roots will allow the plant to establish faster. Large stubby roots that have been cut by digging would indicate that the plant might not have an adequate root system to support it at the new site.

- ☆ Make sure container-grown plants are not "potted" prior to sale. Container-grown plants should have a fine network of roots where the root ball contacts the container. Potted plants often have evidence of larger roots cut at the surface of the root ball. Plants that are dug from the field and sold in containers should be labeled as such. They should have most of the properties of the balled and burlapped plants.
- ☆ The root ball on all dug material should be firm and solid, and the root covering should be in good condition. Observe whether the root covering is burlap or plastic burlap. Plastic burlap should be completely removed before planting.
- ☆ Make sure the crown and root system are in balance on dug plants. Trees with a trunk caliper (diameter) of 1 1/2 inches at a point 6 inches above the root flare should have a root ball no smaller than 20 inches in diameter, and should be between a minimum of 10 feet tall and a maximum of 13 feet tall. Trees with a 2-inch caliper trunk need a root ball no less than 24 inches in diameter and a height between 12 feet and 15 feet. Larger trees with a 3-inch caliper trunk should have a minimum 32-inch-diameter root ball and be between a minimum of 13 feet and a maximum of 18 feet tall. Caliper measurements for trees measuring up to 4 inches in caliper should be taken at 6 inches above the ground. For larger caliper trees, the caliper measurement is taken 12 inches above the ground. Minimum standards for ball sizes exist for all nursery stock based on their size at the time of digging.
- ☆ The root system on container-grown plants should be just at the outer surface of the root ball inside the container. A fine network of healthy, firm, white or tan roots is ideal, but avoid plants with black, slimy, smelly, rotting roots; a dense mass of roots that have circled the inside of the container; or that have begun to extend out the drain holes in the bottom of the container. Dense, circling roots on such plants often result in a girdling root system later in the life of the plant. Such a condition may result in a shortened life and reduced plant quality. Moderate circling roots can be managed by scoring with a sharp knife at planting time to promote new lateral root growth and reduce long-term impact on plant quality.

Purchase plants from a reliable nursery or garden center. Plants from the nursery industry are generally of better quality than those collected from the wild. The roots on field-grown nursery plants are pruned frequently during the production cycle, which results in a more com-pact root system and allows the plants to become established faster. Most nursery-produced plants are given optimum levels of plant nutrients, water, and pest control to create a quality product for the home garden and landscape trade. More tips for selecting quality plants are listed in Table 1.

Nursery plants are available in three forms depending on how they are produced or harvested from the nursery: bare-root, balled and burlapped, or container-grown.

The terms "bare-root" and "balled and burlapped" reflect how plants are offered for sale, while the term "container-grown" describes how the plant was produced prior to sale. The advantages and disadvantages of each form are detailed in table. All three types of plants must be protected from moisture loss once they are harvested and moved from the nursery or garden center.

Table Advantages and disadvantages of bare-root, balled and burlapped, and container-grown nursery stock.

Advantages	Disadvantages
Bare-root	**Bare-root**
• Less expensive • Roots adapt and establish into new planting sites faster • Easy to transport and handle because they weigh less • Large-scale plantings are possible with less expense and easier handling	• Not practical for large plants • Roots may dry out before planting • Small plant sizes make them vulnerable to damage by animals and humans • Must be handled when dormant • Trees may require staking after planting
Balled and burlapped	**Balled and burlapped**
• Greater survival, especially for larger plants and evergreens • Larger specimens are available	• The size and weight of the planting ball makes these harder to handle • Often over 95 percent of the roots are left behind when dug • Availability is limited by season • The texture of the soil of the root ball may dramatically differ from that of the planting area, causing the ball to dry out • Careful handling is required to prevent breaking or damaging the root ball • Before planting, all twine, wires, and treated burlap must be removed or adjusted to ensure long-term survival • The root flare of the tree must be identified to assure planting at the proper depth

Container-grown	Container-grown
• Lighter weight than balled and burlapped plants • Better survival rate than bare-root • Greater availability than bare-root and balled and burlapped • All the roots are moved with the plant at transplanting	• Harder to handle than bare-root stock • Possible root problems including encircling roots and root bound plants because of being in the container too long • The root ball may be grown in a soilless mix and should be completely covered by planting hole soil to prevent drying out of the root ball • May require additional irrigation due to the soilless mix • May be more expensive than field-grown plants

Bare-root

Bare-root plants are harvested from the nursery with no protection over their root systems. They are quite vulnerable to the elements, especially moisture loss. Traditionally, only smaller deciduous trees, shrubs, and some ground covers were offered for sale as bare-root plants. In all in-stances, the plants were small. Recently, city foresters and street tree commissions have adopted bare-root planting of larger street and shade trees for new street tree planting programs. The reduced cost and ease of handling bare-root trees are attractive for street tree programs that have limited funding and that rely on neighborhood volunteers to do the planting. However, the use of bare-root street trees is limited by local availability in the nursery trade and by the fact that not all shade and street trees are adapted to planting as bare-root stock. Bare-root plants should be planted in the early spring or early fall while they are dormant.

Bare-root tree.

Table. A brief list of trees that have been successfully planted bare-root.

Scientific name	Common name
Acer campestre	Hedge Maple
Acer x freemanii	Freeman Maple
Acer platanoides	Norway Maple
Acer pseudoplatanus	Sycamore Maple
Acer rubrum	Red Maple
Acer saccharum	Sugar Maple
Acer x truncatum	Shantung Maple
Amelanchier spp.	Serviceberry
Cladrastis kentukea	Yellowwood
Cornus racemosa	Gray Dogwood
Fraxinus spp.	Ash
Gleditisia triacanthos	Thornless Honeylocust
Gymnocladus dioicus	Kentucky Coffeetree
Malus spp.	Crabapple
Platanus x acerifolia	London Plane Tree
Pyrus calleryana	Callery Pear
Quercus bicolor	Swamp White Oak
Quercus robur	English Oak
Quercus rubra	Northern Red Oak
Robinia pseudoacacia	Black Locust
Sorbus aucuparia	European Mountain Ash
Syringa reticulata	Japanese Tree Lilac
Tilia spp.	Linden
Ulmus spp.	Elm

Balled and burlapped

A balled and burlapped plant is dug from the field with a protective ball of soil over the root system. The ball is held intact with a wrap of burlap, a wire cage, or other support material. All ornamental trees and shrubs, regardless of their size, can be trans-planted as balled and burlapped plants. The size of the tree directly influences the size of the soil ball required to successfully dig and transplant the plant. The larger the soil ball, the heavier and more difficult the task of digging, trans-porting, and planting. The American Nursery and Landscape Association has adopted a series of standards for nursery stock, including minimum ball sizes for balled and burlapped trees and shrubs. Table above outlines the standards, but for more information talk to your nursery or county extension agent. Balled and burlapped plants are most safely transplanted when they are not actively growing. They are available in most nurseries in the spring and early fall.

Balled and burlapped tree

Container-grown

Container-grown plants spend their entire production cycle in a container, so there is no damage to the root system at harvest or sale. Plants grown in containers can be successfully planted any time the soil is not frozen, because none of the roots have been damaged from digging. However, successful planting is more probable in the spring or early fall planting periods. There is a limit to the size of plants grown in a container, but nurseries are using larger containers all the time.

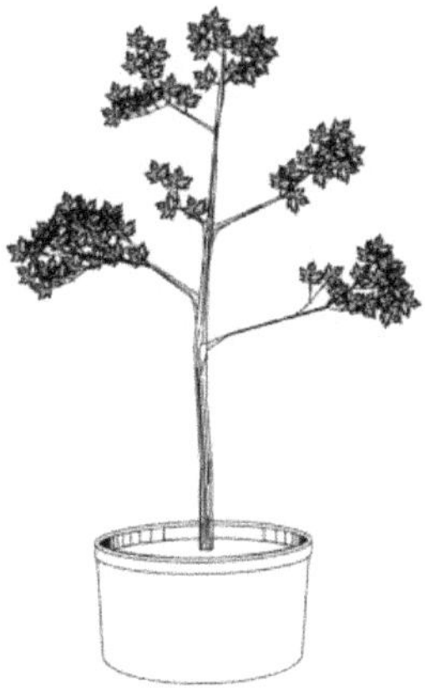

Container-grown tree

3.5.3 Care of Plants before Planting

Protect the foliage, branches, and roots of all plants while they are being transported from the nursery or garden center to the planting site. Always handle or carry plants by their root balls, not their trunks or stems. If you allow the tops of the plants to extend from a vehicle, the foliage and branches can lose valuable moisture. Before you leave the nursery or garden center, the branches (canopy) of trees and large shrubs should be carefully wrapped in burlap or a tarp and secured with rope to reduce moisture stress. After you cover the top of larger plants, they should be laid horizontally in the bed of a truck or trailer and tied down for the ride. Large plants should not be trans-ported while standing straight up in the back of a truck or trailer. Smaller plants can be carried inside the vehicle or bunched together and securely covered with a tarp or burlap in the back of a truck or trailer. The entire plant should be protected against loss of moisture at all times before it is planted.

If your plants are delivered by express or freight, remove them from the package as soon as they arrive. Do not remove the covering from the roots, whether they are balled or bare-root. Make sure the material around the root system is moist—plants are often shipped on the dry side to reduce transportation costs. Keep the root system moist at all times to prevent the roots from drying.

Plant the materials as quickly as possible after they are brought to the planting site. If you have to wait, protect them from exposure to drying wind and sun by keeping them in a sheltered, shaded area. The soil around the roots of balled and burlapped plants should be kept moist with periodic sprinkling. Avoid soaking the root ball, because this adds extra weight and may cause the soil to become "fluid" and move away from the root system. A layer of plastic around the root ball will also help to retard moisture loss. Do not enclose the entire plant in plastic; heat buildup inside the plastic may injure the plant.

Container-grown plants will survive quite a long time before they are planted since they are an entire plant in a single unit. They also should be watered regularly to maintain a uniform level of moisture in the root system. If they are placed in a protected area, they will do very well until planted.

Bare-rooted plants are the most fragile and need considerable care before they are planted. If they will be planted within a few days after arrival, simply keep the packing material around the roots moist. If the planting time will extend beyond this period, keep their root systems moist by temporarily planting them (heeling-in) in moist sand, sawdust, peat moss, or soil. Larger plants should have their roots covered with moist peat and burlap. Plastic sheeting will also help reduce excessive moisture loss from the root system.

3.5.4 Pruning before Planting

The principles of "preplant" pruning recommended for ornamental plants have changed markedly over the years. All pruning concepts for ornamentals differ considerably from the pruning needs of fruit trees. The pruning needs of new ornamental plants should be confined to cosmetic purposes only. The concept of cutting the top growth back to balance the loss of roots during digging in the nursery has been proven wrong. Research indicates that allowing all the leaves to remain in the plant canopy at planting will speed the development of a new and stronger root system to support the plant at the site.

Cosmetic removal of damaged stems and branches should be done at planting. It is often easier to make a few minor cuts in the tops of a taller specimen while it is on its side next to the hole. Some cosmetic pruning can also be done to maintain the natural form of the plant, whether it is bare-root, balled and burlapped, or container-grown.

Pay close attention to the condition of the roots on bare-root material before planting. Damaged or broken roots on bare-root plants should be cut cleanly to remove any ragged edges that result from digging in the nursery. A new, clean cut will close better after planting. Also consider removing excessively long roots that may not fit into the planned size of the hole. Roots should never be bent to fit the hole size or shape: either shorten the root, or increase the size of the hole.

3.5.5 Soil Modification before Planting

Most plants grow best in a well-drained, rich garden loam. Unfortunately, most home properties do not contain ideal soil. Construction practices often remove much of the quality topsoil while leaving only the poorer, slowly drained subsoil as a landscape-growing medium.

If the existing soil drains well and does not contain great amounts of either clay or sand, it will probably support most plant growth with little or no further modification. Research suggests that optimum plant establishment is possible when plants are set into existing soils with good internal water drainage. The same research also indicates that incorporating large amounts of organic matter into the backfill of individual specimen trees and shrubs set into a single hole may be harmful to their survival. The higher amount of organic matter often holds too much water around the root system, and the plant smothers from lack of soil oxygen.

In most Pennsylvania soils, some soil modification may be beneficial before planting individual trees or specimen shrubs if the modifying materials are kept to a maximum of 10 percent of the total volume of backfill material placed around the root system. Common materials used for modifying soils are peat moss, well-rotted sawdust, composted materials such as yard waste, and cow manure. Incorporating any of these materials will improve the structure and water-holding capacity of the backfill soil. Never incorporate sand alone in a heavy or clay soil; the resulting mixture will be very similar to concrete when the soil is wet or dry. All organic matter soil amendments should be mixed uniformly into the backfill.

It is difficult to alter or modify the soil fertility for individual specimen plants. Research indicates that good physical soil quality can be more important to initial plant establishment than fertility. The developing root system needs optimum levels of air and water to become established at the site. All but the poorest soils will generally have enough fertility to aid in the new root growth on these plants.

Testing the soil before selecting plants can prevent the all-too-common problem of choosing a plant or group of plants that have a soil pH or nutrient requirement that cannot be easily met. A soil test mailing kit can be purchased from the Penn State Extension office in your county. Soil test data will provide valuable information on adjusting soil pH and fertility to increase the success of any planting program.

Soil pH requirements vary from plant to plant. If the soil test indicates that the soil pH differs greatly from the needs of the plant, you may wish to consider alternate plants that will fit the existing pH levels of the site. It often takes several years to raise or lower the soil pH more than one pH unit. Planting beds that have had their soil pH artificially adjusted will require constant monitoring and maintenance of the pH.

As a further word of caution, attempts to make large, rapid pH adjustments can alter the soil chemistry to a point where either nutrient deficiencies or toxicity may be produced. If you choose to adjust the pH, do it while preparing the soil before planting. Incorporate proper amounts of lime or soil-acidifying materials required to adjust the pH to the crop needs as noted on the soil test report.

Traditionally, phosphorus (superphosphate) at the rate of 1/4 cup thoroughly mixed with each bushel of backfill or 2/3 cup of bone meal has been recommended to aid in root growth after transplanting. Research has not supported this claim; how-ever, Pennsylvania soils that are low in phosphorus can benefit from such an amendment. The other two major plant nutrients, nitrogen and potassium (potash), can be added later to the surrounding soil as the plant establishes itself. Phosphorus does not move easily through the soil and should be incorporated into the root zone, while nitrogen and potash materials are more water-soluble and will leach into the root area.

3.5.6 Preparing the Planting Hole

Specimen Plants

Specimen plants are generally planted alone, apart from other similar plants. Planting them usually involves digging a single hole in the lawn or mulched bed.

Such specimens include most of the shade trees, and many small flowering trees or large-maturing shrubs.

When digging a single hole, it is essential to make it deep and wide enough to accommodate the roots of the plant, whether it is bare-root, balled and burlapped, or container-grown. A hole that is 1 to 2 feet wider than the diameter of the root ball, but no deeper than the root ball, will allow ample room for good backfill soil around the roots. Research indicates that larger specimen shade trees and most flowering trees will establish better when set on the firm soil in the bottom of the planting hole. Do not loosen the bottom of the hole or add topsoil below the root ball of larger plants. The added weight of such plants often results in settling of the root ball and possible root damage.

When digging the hole for individual plants, discard any existing sod or place it in a compost pile. Place the good topsoil in one pile next to the hole and the poorer subsoil in another pile. Do not use sod or any other materials in the bottom of the planting hole. The root system of the plants should only be exposed to quality backfill soil. If the subsoil and/or topsoil is heavy clay or otherwise extremely poor, you may wish to amend the backfill soil with organic matter. (See section on soil modification.) With proper soil preparation, much of the native soil dug from the planting hole may be used in the backfill operation.

Shrub and Ground Cover Beds

When planting shrubs and ground covers in groups for a specific landscape effect, it is often easier to consider the entire area to be planted as one single hole. If the bed can be worked up to a depth of 10 to 12 inches, you will have most of the labor for planting done. When the root system of larger shrubs requires a hole deeper than 12 inches, you will have to apply the same principles used for specimen plants.

Working the soil in the entire area will improve the root development for all the plants. Root penetration will be much easier and faster when each individual shrub or ground cover plant does not have to establish itself in small holes opened in the native soil of the site. In addition, any needed soil amendments can be incorporated more easily and mixed with all the soil available to the roots.

If the soil at the planting site is very shallow or poor, you may wish to consider replacing it with better quality topsoil. When this is necessary, you should remove the soil to a depth of 12 to 15 inches and blend the new topsoil with the existing topsoil to aerate a transition zone between the existing soil and the new soil. For marginal cases of poor-quality soil, it may be more practical to remove some of the soil, mix it with good topsoil and organic matter, then restore the original level of the soil in the bed.

When total soil removal from the bed area is not practical, it may be helpful to consider incorporating raised planting beds into the area if the proposed landscape design will accommodate such modification. Raised beds use better topsoil as the growing medium, which is placed on top of the existing site soil. Some topsoil should be incorporated into the existing site soil to provide a transition between the poor and better growing soil. Soil depths in a raised bed should be adequate to support all the roots of the plants used. The edges of the raised beds should also extend horizontally far enough to provide for the growth of the roots over time.

3.5.7 General Planting Recommendations

Each plant form—bare-root, balled and burlapped, and container-grown—has specific requirements to consider when planting, and these will be discussed in further detail for each plant form. However, some planting procedures are universal across the plant forms. These procedures are outlined below.

- ☆ The sides of the planting hole should be tapered, with the opening being wider than the bottom and large enough to accommodate the roots and soil ball.
- ☆ The sides of the planting hole should be roughened or scored with a shovel or pickax to break up any smooth surface or glazing created from the backward pressure of the shovel while digging the hole. This polished or compressed layer of soil around the sides of the planting hole will often be dense enough to slow the outward movement of new roots into the soil surrounding the hole.
- ☆ Do not dig too deep—the root flare or collar of the tree should be at the final soil grade or slightly higher. The root collar is the juncture between the stem of the plant and the root system. This point is often associated with a gentle flare when the stem merges into the root system.
- ☆ Before backfilling, make sure the tree or shrub is properly positioned and upright in the planting hole. The plant may require some support while you backfill.
- ☆ Gradually add backfill soil over and among the roots of bare-root plants, and firmly pack soil against the root ball of balled and burlapped and container-grown plants to eliminate air pockets.
- ☆ Continue filling the planting hole and firming the soil until the hole is about half full. At this point, fill the hole with a gentle stream of water and allow it to soak into the soil to settle it and remove air pockets. Avoid a strong stream of water, which may puddle the soil and damage the smaller pores

needed for soil aeration for the roots. When the water has drained, continue filling the hole with soil as recommended under the specific planting system below.

- ☆ Once the hole is filled as directed and all specific planting directions are met, provide a ring of soil several inches high around the edge of the planting hole to hold moisture and fill again with water.

3.5.8 Special Planting Requirements for Bare-Root Plants

The hole for bare-root plants should be large enough to permit the roots to spread out in a normal manner without cramping or twisting. If the roots are curled around the side of the hole, they will tend to continue to grow in that direction and not extend out into the surrounding firmer soil to anchor the plant in the ground. Either trim the longest roots to fit the hole or increase the size and shape of the hole to accommodate the existing root system.

The hole should be deep enough to allow for the natural existing down-ward spread of the root system. If the subsoil is very poor or heavy clay, breaking up the bottom of the hole to allow for better drainage may help. Punching holes into the bottom of the planting pit itself will create escape channels for water to drain away from the root system.

A firmed cone of soil should be placed in the bottom of the hole at a height that will allow the plant to sit at the same height it grew in the nursery. The root collar, flare, or point where the root system and plant stem join should be at the final soil grade or slightly higher, but never lower for ornamental plants.

3.5.9 Special Planting Requirements for Balled and Burlapped Plants

Holes for planting should be prepared as described earlier (1 to 2 feet wider than the diameter of the root ball, and no deeper than the depth of the root ball) to provide adequate space for backfill soil around the root system. If the soil quality at the site is poor, it is advisable to increase the diameter of the hole by 50 percent to allow for a greater amount of better, less compact soil around the root ball.

Before completing the hole, check the root ball on the plant to make certain that the root collar or flare is actually at the top of the root ball and not covered by several inches of soil. At planting, the root collar or root flare should be at the level of the final grade or even slightly higher.

If the root collar is covered with a layer of soil inside the root ball, determine the height of the root ball from the collar to the base of the ball. This measurement should be the actual depth of the planting hole to assure that the root system is at the proper depth in the soil. After the plant has been set in the hole, you should remove this excess layer of soil from the top of the root ball to expose the root collar.

Do not remove the burlap before the plant is set in the planting hole. The burlap will decompose in the soil and can be planted right along with the plant. When the water has drained through the soil, continue filling the hole with soil until it is about 3/4 full.

Once the hole is 3/4 full with soil, loosen the burlap and cord around the top of the root ball, if you have not already done so. The excess exposed burlap and cord can be cut away and discarded, or folded down along the side of the uncovered root ball. Knotted burlap and cord around the base of the plant break down slowly and may interfere with the growth of the plant. Cut and remove all rope used to secure the ball and burlap to the trunk of the tree. Do not allow the burlap to extend above the soil surface, because it will act like a wick and draw needed soil moisture away from the root ball below.

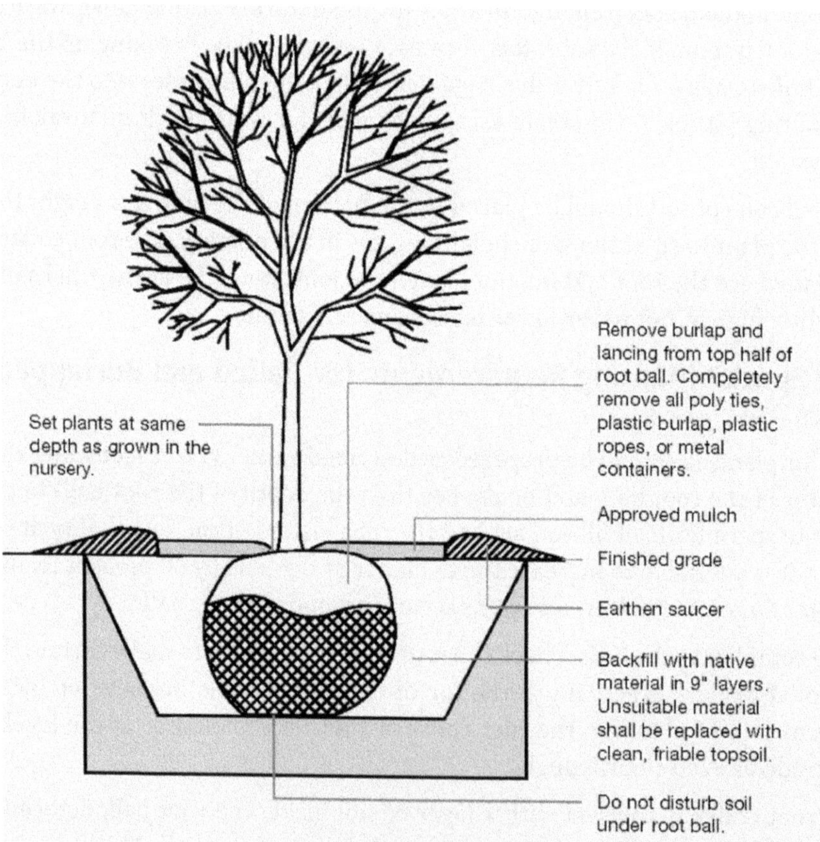

Planting diagram.

Treated and Synthetic Root Ball Coverings

Occasionally balled and burlapped plants are sold with rot-resistant materials covering the root ball. This allows the dealer to hold the plants longer in the sales area and retain a quality root ball. If you are aware of this, you can handle the plants accordingly at home. Treated burlap will usually rot and decompose once it is planted in the soil, but at a slower rate than untreated burlap.

Problems can occur, however, with some plastic or plasticlike fabrics that will not decompose when planted into the soil with the root ball. Many of these materials are so tightly woven that the fibers in the covering cannot be pushed apart by the elongating roots. In this situation, the emerging root system is incapable of expanding into the backfill soil and the plant slowly declines, no matter what steps are taken to promote vigor.

To test for synthetic or plasticlike root coverings, light a match and touch several ends of the burlap. Plastic or synthetic coverings will melt when exposed to a flame. The organic, natural burlap that can be planted with the root ball will burn to ash in a flame.

All synthetic root ball coverings should be removed before the hole is backfilled, but only after the plant has been set in the hole. A small amount of backfill soil may be added around the bottom of the ball, or the sides of the hole can be collapsed to widen the hole and stabilize the plant during the removal steps. Loosen the covering and cut away as much as possible before backfilling begins. The small disk of material that will remain under the base of the root ball should not be a factor in the future establishment and growth of the plant. With the covering removed, you will have to be careful not to crack the soil ball or cause it to drop away from the roots before the entire planting operation has been completed.

Wire Baskets

Machine-dug trees are sometimes placed into a wire basket lined with burlap to protect and support the soil ball during shipping and planting. Once in the basket, the burlap covering is wrapped and secured around the ball in a similar fashion to balled and burlapped trees. The size of the root ball and possibility of damaging the root system makes removal of the basket difficult, dangerous, and unnecessary both before and after planting. Depending on the type of wire used in manufacturing the basket, it is unlikely that the wire will rust away quickly. The intact basket can restrict root growth and girdle roots as they grow out from the root ball. As a compromise, and to ensure that the developing roots are not restricted, the upper half of the basket should be removed using heavy-duty wire cutters before the planting process is completed. Once the upper half of the basket is re-moved, the burlap can be handled as described earlier.

3.5.10 Special Planting Requirements for Container-Grown Plants

When planting container-grown plants, follow the same preparation in hole size and planting depth as you would for balled and burlapped plants. The container has to be removed from this type of plant before it is set in the ground. Even some of the paper materials used for containers break down slowly enough to restrict optimum root development in the soil. Any circling roots in this type of container will continue in the same direction if the container is set with the plant. Of course, plastic or metal materials would be even more harmful to root expansion. Remove the container by either cutting it away or by inverting and knocking the edge of the container sharply and lifting the container off of the root ball.

One additional operation is needed before the plant is set in the planting hole. Take a sharp knife or digging spade and cut an X mark across the bottom of the root ball. Make the cut about one-quarter the depth of the entire root ball. If there is a tight network of roots around the outer edge of the root ball, you should also make several vertical cuts in the root ball from top to bottom. The vertical cuts do not have to be deep, just enough to cut the pattern of the surface roots. By making these cuts, you will encourage the new developing root system of the plant to expand outward into the backfill soil as shown. If the existing pattern of the roots in the container is not disturbed, the roots frequently tend to grow in the same circular pattern established in the container. The developing root system could strangle itself and cause the plant to decline after a few growing seasons.

Scoring the root ball of a container-grown plant.

3.5.11 Guying

Research findings suggest that trees and larger shrubs that are allowed to flex in the wind after planting tend to have stronger stems and trunks. The regular movement of the stems builds a stronger cell system internally in the trunk while it is in the nursery. However, there is also the possibility that such plants will tip or lean slightly prior to sale. Minor adjustments to the position of the plant in the planting hole can alleviate minor twists or curves in the trunk or stem.

In a permanent landscape setting, newly planted trees and larger shrubs should not require staking or guy wires to keep them upright unless the trees were planted bare-root or are planted in an area that is open, windy, or has heavy pedestrian traffic. The dense canopy on shade trees and larger shrubs may offer enough wind resistance to cause some tipping or movement of the plant in the planting hole. Staking holds the plant firmly in place, and allows for faster root development.

As a compromise to staking, it is recommended that the support material be loose enough to allow for some movement of the stem or trunk in the wind. Such support will provide the desired flexing of the stem for the development of added strength, but will not allow any serious movement of the root system that would alter the plant's landscape effectiveness.

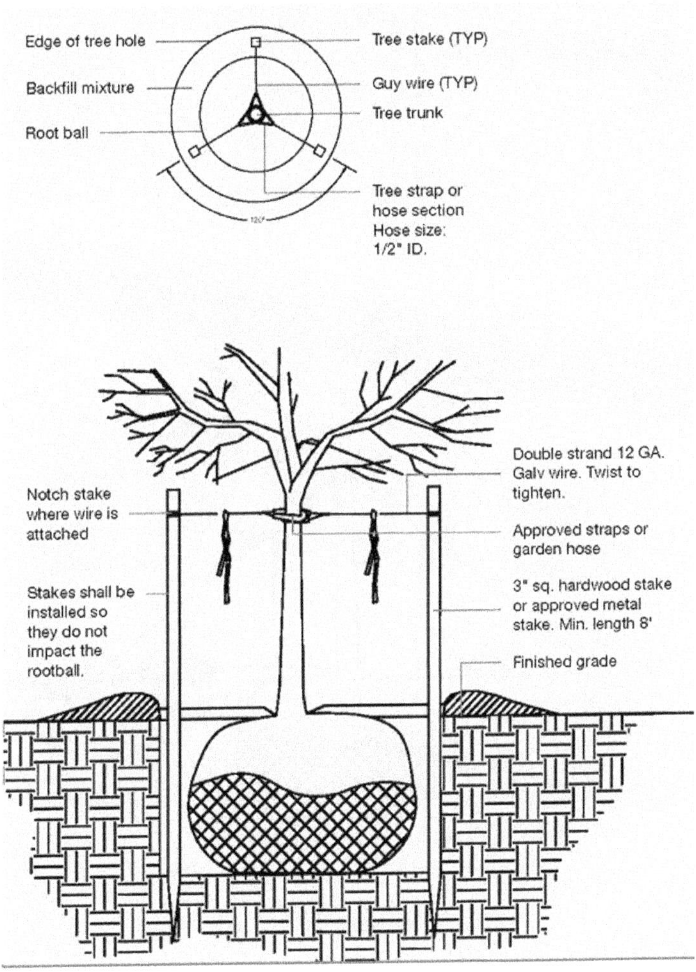

Tree guying diagram for small trees.

There are several methods of staking and guying. Trees 2 inches in caliper or less can be supported by a 2-by-2 stake extending up to the first whorl of branches on the trees. Usually a stake at least 8 feet long is needed. The stake should be driven into the planting hole along the side of the root ball after the plant is set, but before soil has been placed around the root system. The tree can be attached to the stake by a single loop of rope, tree strap, or a wire running through a piece of hose. Cross the hose or rope between the stake and the tree to prevent chafing the bark. Never wrap a wire or other support tightly around a plant stem. As the stem grows and expands in diameter it will grow over the material and strangle the plant.

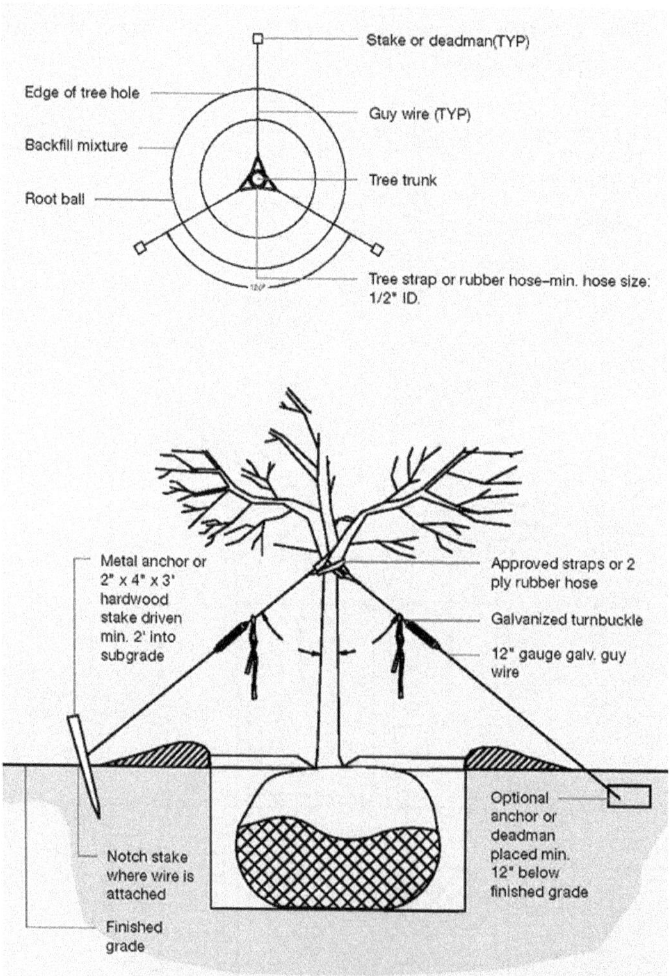

Tree guying diagram for large trees.

Another method of support for trees larger than 2 inches in caliper is to guy the tree with wires. Three guy wires are attached to the tree at equal distances around the trunk. The wire is run through a piece of hose encircling the trunk at a crotch. Where wires are used, anchor stakes should be driven in the ground at a 450 angle, and 1/2 the height of the tree away from the trunk.

As soon as the tree has become established, the support stakes and guy wires can be removed. They should never be left on longer than a full year after planting.

3.5.12 Wrapping

Protecting the trunks and branches of newly planted deciduous trees with tree wrap has become a questionable practice. It has often been suggested that such wrap harbors insect pests that may bore into the trunk. Tree wrap applied in the nursery or garden center before the plants are sold often covers serious defects, such as pruning wounds, bark damage, cankers, or insect injury on the trunk.

Tree wraps do have an advantage when applied and used properly: they will protect the tree trunk and bark from damage during transit from the nursery and while handling the plant before planting. Once the plant is set in the hole, the wrap can be safely removed.

If you decide to use tree wrap, select one of the various commercial paper products or cut 2- to 3-inch-wide strips of burlap. Start the wrap on the larger branches and work down the.trunk to the ground. Overlap each spiral turn about 1/2 the width of the strip so that a double thickness is applied. Bind the wrap loosely with cord or string, starting at the base of the trunk and winding upward to the opposite direction of the wrap.

If you decide to keep the tree wrap on for a short period of time after planting, check the cord or string from time to time to make sure it is not cutting into the bark. The trunk of the tree can begin growing in diameter shortly after it is set in the ground, so careful attention is needed. If it appears that the cord is cutting into the bark, either loosen it slightly or remove it from the trunk. Tree wraps can be used for temporary protection on trees with thin or tender bark during the winter months to prevent frost cracking or sunscald. The wrap should be applied starting at the bottom with each turn slightly overlapping the top edge of the previous turn. This assures that the lower edge of the wrap remains open, allowing moisture to escape from under the wrap on warm days.

Wrapping is not practical on most small shrubs and evergreens; however, they still lose valuable moisture from their foliage and stems. Depending on their location and size, you may wish to consider wind breaks such as snow fencing or burlap barriers, or using an antitranspirant to protect your shrubs and evergreens from moisture loss. These materials are available at most garden centers. Antitranspirants form a protective covering over the foliage and stems to prevent excessive moisture loss.

They have been helpful in hot weather and at exposed locations where moisture loss is above normal. Follow the product's directions for proper application and use.

3.5.13 Mulching

Another useful practice to help conserve moisture for the plant's use is mulching. By placing a layer of mulch over the root system and surrounding soil, you will prevent evaporation of moisture from the soil surface. The mulch will also help to stabilize soil temperatures, control weed growth, and reduce bark damage from mowers and weed whips. Mulching can be used to advantage under specimen plants as well as in shrub beds after they are planted. Mulch around plants set in early fall will prevent rapid freezing and thawing of the soil during winter and early spring.

A variety of materials will function as mulch. However, for the best appearance you might consider such materials as shredded tree bark, aged or composted wood chips, or commercial bark chips. Well-aged compost may also be a possibility if its particle size and appearance are not objectionable.

Proper application of the mulch is important. To be effective, a 2- to 3-inch layer of the material should be applied to the soil surface. Any less than that will not control weeds or reduce moisture loss; more than 4 inches will be of little advantage for weed/moisture control, and may even be harmful to the root system. Deep layers of mulch tend to smother roots by excluding air from the soil.

Organic mulch material should not come in contact with the stem of the plant. Simply scoop the mulch away with your hand once the plant or bed area has been covered. Mulch piled against stems and trunks creates several future health problems for trees and shrubs, including keeping the bark continuously wet, promoting disease and rapid decay of both the bark and wood of the plant, increasing the plant's susceptibility to temperature extremes in the winter, and rodent damage. For the safety of the plant, keep all mulch material from touching the bark. Inorganic materials like stone chips, gravel, or river rock will not hold moisture, but may be abrasive against the bark, which will also injure the stem and lead to plant damage.

Synthetic fabric and plastic sheeting can also serve as a mulching material in certain applications. They are probably most effective in larger areas that have not been planted as shrub beds or where the area will not be disturbed for long periods of time. These products are used under layers of other mulch and function by excluding light from the soil surface, thereby preventing most weed growth.

There are several disadvantages with these products when used directly over the root systems of many plants. Plastic sheeting in particular tends to smother the root system by preventing the exchange of soil gases with the atmosphere as well as the entrance of oxygen into the soil. Plastic sheeting will also limit the entrance of moisture into the soil either from rainfall or irrigation, keeping the soil slightly drier. Likewise, in very wet soils the plastic sheeting may keep the soil too wet and reduce the amount of soil aeration available to the root systems of the plants under the mulch. If you already have extensive areas mulched with plastic sheeting, it would be advisable to puncture the plastic in a number of places with a spading fork to allow for water penetration and the exchange of oxygen and soil gases. The small punctures will not compromise the weed-controlling ability of the plastic. Small collars of plastic under decorative mulch layers around the base of larger trees do not generally cause damage to the plants.

Landscape fabrics, on the other hand, do allow for the movement of water into the soil and the exchange of oxygen and soil gases with the atmosphere. When a layer of organic or inorganic mulch material covers these fabrics, it is possible for weeds to germinate in the upper layer of mulch. As the roots penetrate down-ward, they pass through the fabric and firmly anchor themselves in the soil layer below. When this occurs it is very difficult to pull the weeds without disturbing the entire surface layer of the mulched bed.

Under certain situations, it has also been observed that plant roots tend to form at the interface between the soil surface and landscape fabric instead of penetrating deeply into the soil where they will be able to survive adverse conditions. The zone under the fabric contains optimum conditions for root growth: moisture, oxygen, and darkness. Roots will grow and survive in this location as long as conditions are optimum. However, during dry periods this interface layer will dry out to a point where the root system may be injured. Landscape fabrics are probably most useful on sloping sites that do not contain plants, and where the surface mulch serves some aesthetic purpose in the design. The physical properties of landscape fabrics do allow them to bind well with most organic mulch materials and prevent mulch movement on sloping areas.

3.5.14 Watering

Watering newly set plants after the initial planting is essential for survival. The amount of water needed and when to apply it will depend upon rainfall, moisture-holding capacity of the soil, drainage, and the type and texture of the soil. Watering at 5- to 7-day intervals during the first growing season is usually advisable unless the site dries out faster. Light watering will keep the soil surface wet and encourage undesirable shallow root development. A rule of thumb is to provide 1 to 2 gallons of water per inch of trunk caliper. When watering newly set plants, it is very important to soak the root ball sufficiently and to soak the backfill soil to the depth of the roots.

This is best accomplished with a slow, soaking watering. Watering containers such as a tree irrigator bag, a 20-gallon green perforated bag that wraps around the tree (Figure 9), or gallon water jugs with holes punctured in the bottom can be spaced around the trunk above the root ball.

It is equally important not to overwater newly set plants in sites where the natural soil drainage is slow. It is possible to fill the planting hole and backfill soil with water and drive out all soil oxygen if water is applied faster than it can drain away from the root system.

As the plants become established in the planting site, it is also advisable to extend the area of water application away from the original planting hole. The surrounding soil needs to be kept moist in order to support the new roots expanding beyond the backfill material. A deep root system encouraged by keeping the soil moist around the entire root area will generally produce a better and stronger plant that is able to withstand dry periods later.

Rapid water loss through the foliage on hot, humid summer days often results in wilting and death for many newly set plants. Although the soil is sufficiently supplied with moisture, this condition can occur if the roots cannot function well enough to make up the transpiration loss. Misting the tops of newly set plants will help avoid this condition, but should not be looked upon as a replacement for optimum soil watering practices.

The antitranspirants mentioned earlier can also be used with some success in controlling excessive moisture loss for brief periods shortly after plants are set. As the leaf expands, the material becomes less effective, so repeated applications may be necessary. Follow the package directions for proper application of any material used on the plants.

Tree irrigator bag around a newly planted tree.

3.5.15 Care after Planting

Attention to the following items will help ensure proper establishment of your newly set plants, whether they are specimen, shade, or evergreen trees; small flowering trees; or shrubs and ground covers set in a landscape border.

Maintain optimum moisture in the soil at all times. Do not overwater so the soil becomes saturated. Slightly dry roots will go dormant and come back with moisture, but roots injured from overwatering are gone forever and weaken the plant.

On larger plants that are staked or guyed, maintain solid support for the first year or until the plant has established during its first growing season.

Keep the mulch layer at an optimum depth on the soil surface under the plants. Keep all mulch material from touching the plant stems.

No fertilization is needed for at least the first growing season after planting. Fertilization can start the second season after planting for most landscape plants and the third season for larger trees. Fertilizer application rates should follow a soil test report recommendation relative to the plants chosen. However, if a test report is not available, a light application of a slow-release fertilizer in the fall of the year after leaf color is sufficient. As soon as the plants are dormant, apply a fertilizer with a nitrogen (N), phosphorus (P) and potassium (K) ratio of 3:1:1 or 3:1:2. Commercially available fertilizers with analysis of 18-6-12, 24-8-16, or 15-5-5 are good examples. The slow-release fertilizer should be applied at a rate equivalent to 2 to 4 pounds of N per 1,000 square feet. For an 18-6-12 fertilizer this would equate to 1 to 2 pounds of fertilizer per each 100 square feet of surface area under the plants. Keep the fertilizer about a foot away from the stems of shrubs and the trunks of larger trees. Most of the active root system will be out away from the main trunk. Water the plants well after the application of fertilizer.

3.6 CARE OF ORNAMENTAL PLANTS IN LANDSCAPE

3.6.1 Introduction

Most established ornamental plants in the landscape require care to stay healthy and attractive. Regular fertilization, pruning, watering, mulching and pest control are all part of a good landscape management program.

Some plants, such as annual flowers, roses, red-tip photinia and English laurel, are more demanding than others. Annual flowers and roses must be kept well-supplied with nutrients and water at all times, pruned and groomed routinely, and monitored regularly for pests if they are to flower abundantly. Red-tip photinia and English laurel are very susceptible to leaf spot diseases and demand more care.

On the other hand, plants such as ornamental grasses, hollies and junipers, when properly planted on a good site, require little care once established and are considered good low-maintenance plants.

Caring for ornamental plants is more difficult when they are not well-suited to the site selected or when they are improperly planted. Azaleas, for instance, prefer a moist, well-drained soil and shade from the mid-afternoon sun. When planted in poorly drained soil or in full sun without the benefit of irrigation, azaleas become stressed, susceptible to attack from insects and diseases, and require more care. Planting too deeply or too shallowly also causes plant problems.

Today, the concept of low-maintenance landscapes is more popular than ever before. The idea of putting less effort into the landscape without sacrificing quality and beauty is very appealing to a growing number of homeowners and clients of landscape professionals. By applying lower amounts of fertilizer, water and pesticides to the landscape, you not only help the environment but also save time and money. New concepts in low-maintenance landscaping, such as Xeriscaping (water-efficient landscaping) and integrated pest management (control of pests through the selective use of insecticides and encouragement of natural enemies), are proving that it is possible to have a beautiful landscape while saving time, effort and money.

A recently planted area, for instance, will generally require more attention than a well-established area. The highly visible public area of the landscape is usually an are where optimum growth is desired at all times, while a secluded, private area requires less maintenance. Once this "zoning" is done and the different levels of plant care are established, landscape maintenance becomes much more efficient and effective.

3.6.2 Fertilization

Fertilization is an important part of landscape maintenance, particularly in urban areas where much of the native topsoil is removed during development and subsoil deficient in essential nutrients becomes the new topsoil.

Unfortunately, fertilization is a rather simple cultural practice made complex and confusing by the wide variety of fertilizer products on the market today -- from "general-purpose" garden fertilizer to specialty products, such as pre-mixed liquid fertilizer concentrates, water-soluble crystals, slow-release fertilizers, azalea/camellia fertilizers and rose specials. Plants generally do not care whether a fertilizer is granulated, liquified, encapsulated, briquetted, pelletized or solubilized. They simply want nutrients in any form they can use.

Fertilizers, however, differ in nutrient content and release duration. The type of fertilizer you select should be based not only on its cost but also on the types of plants being fertilized, the existing nutrient content of the soil, and the type of growth response desired. Liquid or water-soluble fertilizers, for instance, are often used on annuals and herbaceous perennials immediately after transplanting because their nutrients can be absorbed quickly and used by the plant. Woody ornamentals, on the other hand, store food reserves in their roots and do not have an immediate demand for nutrients after transplanting as short-season annuals do. They benefit from slow-release fertilizers that ensure a supply of nutrients as needed.

To determine whether a granular fertilizer has slow-release properties, look at the analysis on the back of the bag. Nitrogen listed in the form of ammoniacal nitrogen indicates that the product has some slow-release property. If the nitrogen is listed as being derived from urea, urea-formaldehyde, IBDU (isobutylenediurea), or sulfur-coated urea, the release duration of the product will be increased. Some granular slow-release fertilizers last six to eight months after application.

	% Nitrogen (N)	Phosphorus (P_2O_5)	% Potash (K_2O)
Blood, dried	13.0	-	-
Bone meal (raw)	3.5	22.0	-
Bone meal (steamed)	2.0	28.0	-
Cottonseed meal	6.6	2.5	1.5
Fish scrap (dried)	9.5	6.0	-
Soybean meal	7.0	1.2	1.5
Horse manure	0.7	0.3	0.6
Cow manure	0.6	0.2	0.6
Pig manure	0.5	0.3	0.5
Sheep manure	0.8	0.3	0.9
Chicken manure	1.1	0.8	0.5
Duck manure	0.6	1.4	0.5

Average nutrient content of various organic fertilizer sources.

Other commonly available slow-release fertilizers on the market include Osmocote granules, Osmocote tablets, Jobe's Spikes, Once, Woodace briquettes, Agriform tablets and Milorganite. These fertilizers generally cost more per pound than general-purpose granular fertilizers such as 10-10-10 or 12-4-8, but they also last longer and don't need to be applied as frequently.

Organic fertilizer sources such as bone meal, cottonseed meal and animal manures can also be used. Table below lists the average nutrient content of several organic fertilizer sources. Compost is another good source of slowly available nutrients.

A soil test, available through your county Extension office, is the best way to determine which fertilizer analysis is best for your soil. As a general guideline, most ornamental plants will benefit from a fertilizer having its primary nutrients (nitrogen, phosphorus and potassium [N-P-K]) in a 3-1-2 or 4-1-2 ratio. A 12-4-8 fertilizer, for instance, is a 3-1-2 ratio, and a 16-4-8 fertilizer is a 4-1-2 ratio. Research shows that phosphorus, the middle number in the analysis, is held be soils and does not leach with rains or irrigation as nitrogen or potassium do; so it is usually needed in lower amounts. On new sites where phosphorus has never been applied, a complete balanced fertilizer such as 8-8-8 or 10-10-10 is sometimes recommended.

Research shows that woody plants actively absorb nutrients from the soil during the growing season and require few nutrients during the dormant winter season as shown. Therefore, apply fertilizer as soon as the plants begin breaking dormancy in the spring, and avoid fertilizing after the first fall frost, which signals plants to begin resting for the winter.

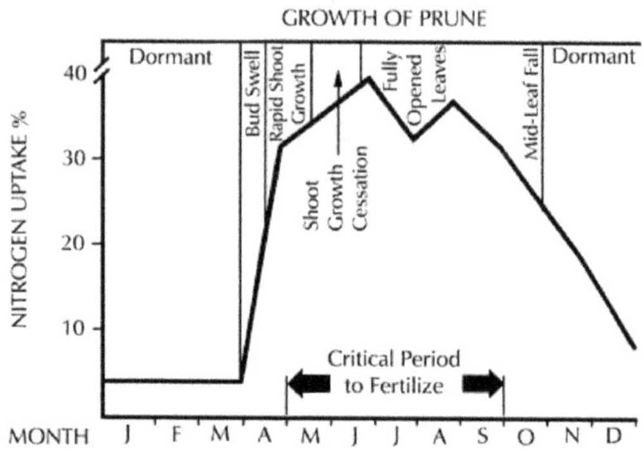

The relationship between nitrate uptake efficiency of non-bearing prune trees and their growth cycle.

The frequency of fertilization depends on the type of plants being fertilized and the type of fertilizer used. If slow-release fertilizers are used, one application should be sufficient for the entire growing season. If general-purpose granular fertilizers are used, two or three applications may be needed, depending on the fertilizer's slow-release properties.

When general-purpose granular fertilizers such as 12-4-8 or 10-10-10 are used on newly-planted ground covers, annuals, herbaceous perennials and roses, light applications made at four- to six-week intervals are recommended.

During periods of limited rainfall or drought, reduce the amount of fertilizer applied and the frequency of application in non-irrigated areas. Fertilizers may injure the roots of ornamentals under drought stress.

Newly-planted ornamental trees and shrubs will benefit from light applications of fertilizer made during the first growing season after transplanting. Uniformly broadcast the fertilizer along the perimeter of the planting hole. Avoid using weed-and-feed fertilizers in the vicinity of newly-planted ornamentals because injury from the herbicide may result.

Newly-planted 1-gallon size plants will respond to 1 level teaspoon of a 12 to 16 percent nitrogen fertilizer or a level tablespoon of an 8 to 10 percent nitrogen fertilizer applied in March, May and July. Small trees, fewer than 4 feet tall, should receive no more than 1 tablespoon of a 12 to 16 percent nitrogen fertilizer two to three times during the first growing season. Larger trees will benefit from 3 to 4 tablespoons of a 12 to 16 percent nitrogen fertilizer. Broadcast fertilizer along the perimeter of the planting hole. Remember that newly trans-planted ornamentals are under stress while they are trying to adapt to their new location and they can be easily injured by over-fertilization.

Table. Recommended fertilization rates for newly planted ornamental plants during the first growing season (use only one of the fertilizers listed at the rate recommended).

	Application rate[1]/plant			Application frequency
Plant type/size	12-4-8	16-4-8	10-10-10	
1-gallon shrubs	1 tsp.	1 tsp.	1 tbsp.	March, May, July
3-gallon shrubs	2 tsp.	2 tsp.	2 tbsp.	March, May, July
5-gallon shrubs	3 tsp.	3 tsp.	3 tbsp.	March, May, July
Trees under 4 feet	1 tbsp.	1 tbsp.	2 tbsp.	March, July
Trees 4 to 6 feet	3 tbsp.	3 tbsp.	5 tbsp.	March, July
Trees 6 to 8 feet	4 tbsp.	4 tbsp.	6 tbsp.	March, July
	Application 100/sq. ft			Application frequency
	12-4-8	16-4-8	10-10-10	
Ground covers, annuals & herbaceous perennials	0.5 lb.	0.5 lb.	1 lb.	Each 4 to 6 weeks

The quantity of fertilizer applied on established ornamentals depends on the analysis of the fertilizer used, the area fertilized and the amount of growth desired. Nitrogen controls vegetative growth, so application rates are based on this primary nutrient. Table below lists suggested application rates for several general-purpose fertilizers.

To increase the application rate, increase the frequency of application, but do not exceed the amount recommended in table for each application. Optimum growth fertilization rates (three to five applications) are usually used on annuals, herbaceous perennials, roses and newly established ground covers to encourage their spread. Otherwise, low rates of fertilizer are recommended, particularly if you desire a lower maintenance landscape. As the application rate of fertilizer increases, so does the amount of new growth, which requires more water, more fertilizer and more pruning.

Table. Recommended application rates of various general-purpose granular fertilizers on established ornamental plants in the landscape.

	Application rate*				
	1000 sq. ft.		100 sq. ft.		10 sq. ft.
Source	pounds	cups	pounds	cups	tablespoons
10-10-10	10.0	20	1	2.0	4.0
8-8-8	12.5	25	0.5	2.5	5.0
13-13-13					
12-3-6	6.0	12	3/4	1.5	3.0
12-4-8					
12-6-6					
16-4-8	6.0	12	0.5	1.0	2.0
4-12-12	25.0	50	2.5	5.0	10.0
5-10-10	20.0	40	2	4.0	8.0

To determine how much fertilizer to apply, first estimate the area to be fertilized. This involves estimating the length and width of a bed in linear feet and multiplying the two numbers to obtain square footage. Few plant beds are perfectly square or rectangular, so square off the rounded areas to simplify your estimate. Trees growing within a bed can be included in the bed estimate or, if they require special fertilization, estimate their canopy area by measuring the distance from the trunk to the drip line or tip of the branches (this is called the radius). Then use the geometric formula for the area of a circle to calculate the area of the canopy: 3.14 x radius^2. For example, if the distance from the main trunk to the drip line of a tree is measured to be 20 feet, the area beneath the canopy is 3.14 x (20 x 20) = 1,256 square feet. See guidelines below for additional recommendations on tree fertilization.

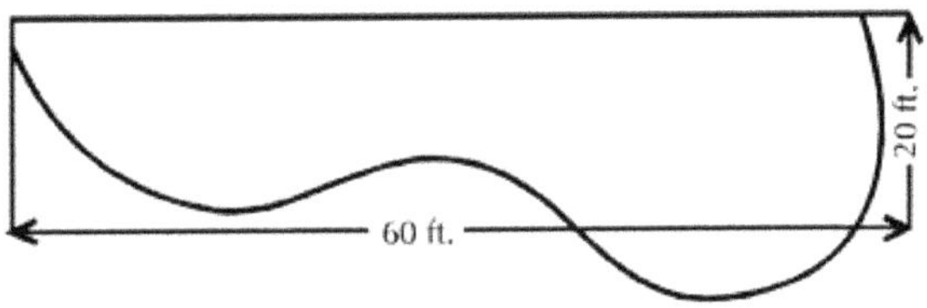

Calculation of square feet in plant bed

3.6.2.1 Guidelines and Precautions when Fertilizing

- ☆ Broadcast fertilizer evenly over the bed area. Make certain the foliage is dry when fertilizer is broadcast over the tops of ornamentals.
- ☆ If fertilizer becomes lodged in the foliage of ornamentals, brush it off before irrigating.
- ☆ It is not necessary to remove the mulch when fertilizing. Broadcast fertilizer on top of the mulch and water it in.
- ☆ Trees growing in turf areas will obtain nutrients from the fertilizer that is applied to the turf grass. Do not apply excess fertilizer to turf in an effort to feed trees; injury to the turf may occur.
- ☆ When fertilizing trees, broadcast the fertilizer over an area extending two to three times the canopy spread if possible. Research has shown that tree roots grow far beyond the canopy spread on established trees.
- ☆ Do not concentrate fertilizer in holes drilled under tree canopy. Research shows that broadcast application results in better growth.
- ☆ An effective technique for fertilizing annual flowers is to place a slow-release fertilizer such as Osmocote in the planting hole directly beneath the plant. This not only provides an even supply of nutrients to the plants, it also decreases the number of weeds when compared to broadcast application.
- ☆ Plants growing in shade generally require less fertilizer than those growing in the sun.
- ☆ Plants growing in sandy soils generally require more frequent fertilization than those in clay soils due to nutrients leaching from sandy soils.
- ☆ Avoid using weed-and-feed lawn fertilizers that contain herbicides near ornamental plants.

3.6.3 Watering

Most ornamental plants in the landscape, once they are established, can go days or even weeks without supplemental irrigation. In fact, over-watering (water too frequently) is a leading cause of problems with ornamentals. Junipers, for example, are extremely drought-tolerant once they are established, but they cannot tolerate extended periods of excess moisture.

Your best guide for determining when to water is the plant itself. Wilting or a pale grayish-green color are the most common symptoms in plants needing water. Certain plants in the landscape -- such as annuals, herbaceous perennials, azaleas and rhododendrons -- need more water than others. By watering only those plants that need water, you not only will save water, time and money, you also avoid watering plants that do not need a lot of moisture.

Drip irrigation saves water in landscape and is ideal for ornamental plants.

To avoid run-off and water loss, apply water slowly to the base of the plant using a hand-held hose, drip or trickle irrigation, micro-sprinklers, or an ooze hose. Do-it-yourself irrigation systems, available from most garden centers, use 30 to 50 percent less water than sprinklers, and they can be attached directly to outdoor faucets.

The amount of water needed by an ornamental plant depends on the type of plant, the soil type, the amount of existing moisture in the soil, and the time of year. As a general rule, 6 gallons of water per 10 square feet of bed area or canopy area will saturate most soils to a depth of about 12 inches (the area containing 80 percent of the roots of most ornamental plants). Because water moves readily within the plant, you do not need to water the entire root zone. Twenty-five percent of the root area can absorb enough water for the entire plant.

The best time to water is at night or in the early morning. As much as 30 percent of the water applied during midday can be lost to evaporation accelerated by the sun.

Hydrogels (water-absorbing polymers), sold in garden centers under several different trade names, are popular products on the market. They absorb several hundred times their weight in water and release it slowly, as needed, to the plant. Research at The University of Georgia shows hydrogels may provide a significant growth benefit to summer annuals in non-irrigated soil.

3.6.4 Mulching

Mulches are a vital part of the Southern landscape. They hold moisture in the soil, help prevent weed growth, inhibit certain soil-borne foliar diseases, and insulate the roots of plants from temperature extremes during summer and winter. Mulch also provides a buffer zone between the turf and woody ornamental plants and helps prevent trunk injury from weed trimmers and lawn mowers.

The best mulch is organic, fine-textured and non-matting. Examples include pine needles (commonly called pine straw in the South), pine bark mini-nuggets, hardwood chips and cypress shavings. Fall leaves are an excellent and economical mulch and add valuable humus back to the soil as they decompose. Pecan hulls, a by-product of the pecan industry, are used successfully as mulch in south Georgia. Grass clippings are not a good source of mulch because they tend to mat down and inhibit the flow of water and nutrients into the soil. They also my introduce weeds into ornamental plantings. Inorganic mulches such as rock, gravel and marble are good soil insulators, but they absorb and re-radiate heat in the landscape, increasing water loss from plants. Limit their use to shady areas of the landscape.

Apply mulches 3 to 5 inches deep in an area extending beyond the canopy of the plant if possible. A small ring of mulch around the trunk provides little benefit to the plant.

Organic mulches such as pine straw or pine bark break down and decompose over time and should be replenished at least once a year.

Avoid placing plastic film under mulches; it prevents water, nutrients and oxygen from reaching the roots of the plant. Geo-textile (landscape fabric) can be used instead of plastic film under mulch. It allows water, nutrients and oxygen to reach the roots of plants and prevents the growth of some weeds. For best results, install landscape fabrics on weed-free ground and avoid getting soil on top of the fabric.

Landscape fabric

3.6.5 Pruning

Ornamental plants in landscapes are pruned for many reasons. We usually prune plants to achieve or maintain a certain size or form. At other times, we prune to remove old, diseased or damaged wood in an effort to promote vigorous new growth, flowering or fruiting.

Two basic pruning techniques are used to prune ornamental plants -- heading and thinning. Heading (also called heading back) refers to the non-selective removal of branches; thinning is the selective removal of branches back to a side branch or main trunk. Shearing is a form of heading.

Thinning results in open canopy, improved air circulation and natural shape; shearing (left) encourages a dense outer canopy and unnatural form.

Thinning is best for most ornamental plants because it opens up the plant canopy, increases air circulation within the plant, and results in a natural growth form. Shearing is frequently done on small-leaf evergreens such as boxwood and Japanese holly. Shearing is an efficient pruning technique, but it causes a thick outer canopy and loss of natural form.

If low-maintenance is your goal, keep shearing to a minimum in the landscape. Once you start shearing plants, you will have to shear them more frequently to keep them looking good. Selective thinning of branches once or twice a season is much less labor-intensive than monthly shearing.

Two ornamental plants in the landscape that are commonly sheared are liriope and ornamental grasses such as pampas grass. Shearing in late winter removes old growth and makes way for new shoots. An efficient way to prune liriope is with the lawn mower set at the highest possible setting.

Prune flowering ornamental plants at the proper time of year; otherwise flower buds might be removed. As a general rule, prune spring-flowering trees and shrubs soon after they bloom. Summer-flowering trees and shrubs bloom on new wood and are best pruned four to six weeks before spring growth begins. Table 4 provides suggested pruning times for several common flowering trees, shrubs and vines.

Some shade and flowering trees tend to bleed or excrete large amounts of sap from pruning wounds made in the early spring. Among these trees are maple, birch, dogwood, beech, elm, willow, flowering plum and flowering cherry. Sap excreted from the tree is not harmful, but it is unsightly. To minimize bleeding, prune these trees after the leaves have matured.

Avoid pruning during the fall or early winter because it may encourage tender new growth that is not sufficiently hardened to resist the winter cold.

Most broadleaf shrubs -- including azaleas, camellias, ligustrum, glossy abelia, nandina, cleyera and crepe myrtle -- will tolerate severe pruning when they overgrow their site or need rejuvenating. This involves cutting back the plant to within 12 inches of ground level. Boxwoods, junipers, pines, cypress, cedar, arborvitae, yews and other narrow-leaf evergreens do not respond favorably when severely pruned and may decline.

The best time to prune severely is when spring growth begins -- mid-March in north Georgia and mid-February in south Georgia. Avoid pruning severely from late fall to mid-winter since it increases the chances of cold injury to the plant.

Table. Suggested pruning time for common flowering trees, shrubs and vines.

Prune after flowering:			
Azalea	Deutzia	Japanese Pieris	Star Magnolia
Beautybush	Dogwood	Lilac	Shrub Honeysuckle
Bigleaf Hydrangea	Doublefile Vibernum	Mockorange	Thunberg Spirea
Bradford Pear	Flowering Almond	Oakleaf Hydrangea	Weigela
Bridalwreath Spirea	Flowering Cherry	Pearlbush	Winter Daphne
Clematis	Flowering Quince	Pyracantha	Wisteria
Climbing Roses	Forsythia	Redbud	Witchhazel
Crabapple	Japanese Kerria	Saucer Magnolia	

Prune before spring growth begins:			
Beautyberry	Floribunda Roses	Japanese Barberry	Rose-of-Sharon (Althea)
Camellia	Fragrant Tea Olive	Japanese Spirea	Sourwood
Chaste Tree (Vitex)	Grandiflora Rosse	Mimosa	Anthony Waterer Spirea
Cranberrybush Viburnum	Glossy Abelia	Nandina	Sweetshrub
Crepe myrtle	Goldenrain Tree		

3.6.5.1 Pruning guidelines

Arborvitae: The exterior foliage on this evergreen can be pruned lightly when it needs shaping. Avoid making major cuts.

Azalea: Prune by thinning after bloom if necessary. Don't prune at all if the plant looks good. Old overgrown plants can be renewed by being cut back close to ground level in February.

Beautyberry: Thin out growth before spring growth occurs. Flowers and fruits appear on new growth.

Birch: Make major cuts when dormant. Light pruning in full leaf will minimize sap flow (bleeding) from wound.

Boxwood: Prune by thinning or shearing almost any time of the year. Avoid heavy pruning because the plant is slow to recover and may decline.

Camellia: Thin out branches after bloom if necessary. Camellias generally require little pruning.

Cherry (Ornamental): Make major cuts in late winter. Use light pruning after bloom to remove suckers or to shape.

Clematis: Some of these plants bloom on old wood and some on new wood, depending on the species. It's best to wait until after bloom to prune this plant. Thin out the old wood. Some vigorous varieties can be pruned within 12 inches of ground level.

Cleyera: Prune by thinning during the growing season to maintain a natural shape.

Cotoneaster: Make thinning cuts to remove old wood and to shape in late winter or early spring.

Crabapple (Flowering): Prune when fully dormant to remove suckers and to produce a desirable shape. Young suckers can be easily removed by hand during the growing season.

Crape Myrtle: This plant flowers on new growth, so prune it when it is dormant. Make thinning and heading cuts to produce a desired form. For shrub forms, prune all branches close to ground level. For tree forms, save one to three of the most vigorous branches for the main trunk(s) and prune off low-growing branches to develop a tree-like growth habit.

Deutzia: Make thinning cuts after flowering if shaping is necessary.

Dogwood: This tree sets blossom buds in late summer. Make major cuts when dormant even though you may sacrifice some blossoms. Otherwise, prune it after flowering.

Elaeagnus: A very vigorous summer grower, it may grow a foot or more a week during the growing season. Thin out long shoots as necessary. Don't try to maintain a formal shape.

Euonymus: The growth habit of this plant ranges from upright shrubs to ground covers. Prune by thinning as necessary any time during the growing season.

Fringetree: Prune by thinning in late winter to achieve desired shape. Birds enjoy the late summer fruit, so avoid pruning after flowering.

Goldenrain Tree: Prune to tree form in late winter.

Hemlock: This plant normally needs no pruning. Light shearing of the outer canopy may be necessary to correct its form. Avoid major cuts.

Holly: There are many different growth habits and forms. Most are evergreen, but some species are deciduous. If plants are prized for berries, prune them in late winter, before spring flowering. If berries are not a concern, thinning (or shearing of small-leaf types) can be done any time during the growing season.

Honeylocust: Maintain desired shape by thinning in late winter.

Hydrangea (Bigleaf, French, Oakleaf): Flower buds form on old wood. Prune after flowering.

Hydrangea (Panicle, Smooth): Blooms form on new wood. Prune when dormant, and remove spent blossoms after flowering.

Jasmine (Winter): Thin out after flowering to maintain desired shape.

Juniper: Maintain shape or eliminate undergrowth of groundcover types by thinning during the growing season. Avoid heavy pruning to old wood because new growth will not occur.

Laurel (English): Prune by thinning as needed during the growing season.

Liriope: Remove old foliage four to six weeks before the spring growing season. On level ground, a lawn mower set to make the highest cut is an efficient way of pruning this plant.

Magnolia (Japanese): Prune to desired shape after flowering.

Magnolia (Southern): Generally requires little pruning. Shape by thinning during the growing season, preferably after bloom.

Maple: Prune in late winter if major cuts are necessary. Light pruning in mid-summer can also be done. Avoid early spring pruning because unsightly sap will flow from the pruning wounds.

Mockorange: Prune after flowering by thinning out old wood. The plant may be cut back to ground level if desired.

Mountain Laurel: Prune lightly by thinning to desired shape after flowering.

Nandina: As plants age, thin out old canes by cutting them back to ground level. Selectively cut back one-third of the remaining canes by half their length to encourage a full-dense canopy. Do this pruning in late winter or after fruiting.

Oak: Prune to desired shape when dormant.

Oleander: Flowers appear on new growth, so prune just prior to spring growth. Thin out old wood and head back top for desired shape and height.

Osmanthus: Shape by thinning during the growing season. The plant responds well to heavy pruning.

Pear (Ornamental): Make major cuts in late winter (when dormant), even though some blossoms may be sacrificed. Lightly prune after flowering if necessary.

Photinia (Red-tip): Prune any time during the growing season. Early spring and late summer pruning result in new growth that turns brilliant red.

Pine (Also spruce, fir): Prune to desired shape by removing all or part of the new growth (called candles) in spring. Avoid pruning into old wood.

Pittosporum: Prune to desired shape any time during the growing season.

Pyracantha: Prune after fruit set to remove non-fruiting wood.

Quince (Flowering): Prune after flowering. Thin out old branches and head back others to desired form and size.

Redbud: Make major cuts in late winter. Light pruning can be done after flowering.

Rhododendron: Prune, if necessary, to achieve desired shape and to increase branching after bloom.

Rose (Hybrid tea, grandiflora, floribunda): Prune in early spring when new growth begins.

Rose (Climbing): After flowering, thin out old canes and head back remaining shoots by about one-third, depending on their vigor.

Rose-of-Sharon: This plant flowers on new growth, so prune it in late winter.

Spirea: Prune by thinning after bloom. Most species respond well to severe pruning.

Trumpetcreeper: Flowers on new growth, so prune it during the dormant season. This plant will tolerate severe pruning.

Viburnum: Prune after flowering or fruit set to thin out oldest non-fruiting wood and to improve shape.

Waxmyrtle: Prune to desired shape during the growing season.

Wisteria: Prune after flowering. This is a very vigorous vine and it can be heavily pruned.

Witchhazel: Prune by thinning after flowering.

General pruning procedure

3.6.6 Weed Control

Weeds in the landscape can be controlled mechanically (by hand, with mulch and landscape fabric) or chemically (with herbicides).

For chemical control, preemergence and post-emergence herbicides are available. Preemergence herbicides control weeds in the early stages of weed seed germination, and post-emergence herbicides are used to kill established weeds. Examples of preemergence herbicides are Dacthal, Surflan and Treflan. Commonly used post-emergence herbicides include Roundup (also called Kleenup or Blot-Out) and Vantage (formerly Poast). For specific detailed information on weed control, consult the Georgia Pest Management Handbook – Homeowner Edition.

3.6.7 Insect and Disease Management

Landscape plants are attacked by a diverse and complex group of insects and diseases. Management of these pests can be achieved by following these guidelines:

- ☆ Many insect and disease problems can be prevented by using plant materials that are seldom attacked or are less prone to pest problems. When buying new plant materials, consider the history of potential insect and disease problems for the plant before your purchase. Your county Extension office can help you with your selection.
- ☆ The best defense for pest problems is a healthy, actively growing, well-maintained plant. Follow proper management programs to maintain plant health.
- ☆ Monitor plants for pest problems on a regular basis, particularly during the growing season. Early detection of a pest problem will provide adequate time for control measures to be taken before the plant is extensively damaged.
- ☆ The correct diagnosis of a plant pest problem is the cornerstone of a successful control program and often requires careful study of the plant and surrounding environment. Diagnosis can often be based on symptoms expressed by the infested plant or by pests seen on the plant.

Diagnosis of some insect and mite problems often requires close examination of the plant. Chlorotic spots on leaves, a rust coloration of the upper leaf surface, or the silvering of the upper leaf surface may indicate the presence of mites, lace bugs or thrips. Curling or distortion of new plant growth, stunting of the plant, the presence of a sticky substance on the upper leaf surface, or a black sooty mold growing on the upper leaf surface indicate the presence of aphids, scale insects or whiteflies. These pests are small and difficult to detect when present in small numbers, so if a plant starts to show any of these symptoms, examine it closely for the presence of these pests.

Another sign of insect problems is the removal of plant tissues. Beetles, caterpillars, bagworms and grasshoppers damage landscape plants by removing the feeding on leaves. When holes are observed in leaves or portions of leaves, or when entire leaves are consumed, examine plants for the presence of one of these pests.

Insects such as borers, twig girders and leaf miners damage plants through their tunneling activities. Holes in the bark of trees, tunnels underneath the bark, tunneling activity in leaf tissue, dead terminal growth on a plant, or girdled twigs indicate an infestation of one of these pests.

The presence of strange plant growths, webbing or silk on plants can also indicate an insect or mite problem. Tent caterpillars and webworms construct silken webs or tents on their host plants. When spider mite populations become large, they cover their host plant with silken webbing. Gall insects often cause their host plants to produce abnormal growths on leaves, stems and twigs.

The presence of insects on ornamental plants does not always mean that control measures need to be taken. Non-pest insects may rest or hide on plants without posing a threat to plants. Light infestations of pest species can be controlled by beneficial insects or environmental factors and no additional control measure are needed. Healthy, actively growing plants can withstand some insect damage with no adverse long-term effects.

Once a pest problem has been properly diagnosed, a control program can be determined. If a chemical treatment is to be used as part of the control strategy, its effectiveness depends on strict adherence to the following:

1. apply the product when the pest is in a susceptible stage
2. select a pesticide that is labeled for control of the insect application on the host plant
3. closely follow label directions concerning dilution rates
4. apply the product as specified on the label
5. apply the product to ensure good coverage of the host plant or target pest.

Chapter 4

PRODUCTION TECHNOLOGY OF CUT FLOWERS

4.1 INTRODUCTION

4.1.1 Scope and Nature of the Flower Industry

The term "Cut flowers" refers to blossoms or flower buds that are cut with branches, leaves and stems. Fresh cut flowers are highly perishable because they maintain only limited life supporting processes by absorbing water through their stems. They are used for bouquet preparation, decorations, vase arrangements; designs for weddings and funerals; gifts on occasions; informal display to beautify home and public places. Cut flowers are the most important floricultural products in the international flower trade. The most traded flower crops include roses, carnations, chrysanthemums, dahlia, gerbera, gladiolus, gypsophila, liatris, nerine, orchids, poinsettias, achillea, anthurium, tulips and lilies. These flowers are cultivated in open fields or protected structure.

The floral industry is one of the higher industries in many developing and underdeveloped countries. Floriculture as an industry began in the late 19th century in the United Kingdom, where flowers were grown on a large scale on the vast estates. The present day floral industry is a dynamic, global, fast-growing industry, which has achieved significant growth rates during the past few decades. In the 1950s, the global flower trade was less than US$3 billion. By 1994, it had grown to US$100 billion. In recent years, the floral industry has grown six percent annually, while the global trade volume in 2003 was US$101.84 billion.

The floral industry essentially consists of three major components: the growers, the wholesalers and the retailers whose businesses are quite intermingled. The recent trends are more towards eliminating the intermediaries, the wholesalers between the growers and the retailers, so that the flowers are made available at considerably low prices.

Some flowers are sent packed flat in boxes. This enables large amounts of flowers to be packed in small spaces like aircraft holds. Other flowers cannot survive for long periods out of water such as orchids, gerberas (gerber daisies) and water lilies. These are either sent with their own sealed water container (called picks) on each stem end - for more expensive or tropical flowers - or are transported in buckets of water (This method of transport in water is often referred to as "Procona"). The latter method extends the life of flowers and reduces labor time as flowers are ready for sale, but obviously also reduces the amount of flowers that can be transported as they are much heavier than dry-packed flowers and hence air transportation charges are higher.

Flowers take a number of routes to the consumer, depending on where they are grown and how they are to be sold. Some growers cut and pack flowers at their nurseries, sending them directly out to the consumer by mail order. Some flowers are sent to packing companies, who grade the flowers and arrange them in bunches for sale to supermarkets or to deliver by mail order. Some flowers are graded and sleeved by the growers and sold at wholesale flower markets; the wholesalers then sell them on to florists who condition and arrange the flowers for the consumer.

Traditionally, the center of flower production has been near their largest consumers: the developed world, where Japan, Western Europe and North America were both major producers and consumers. The major consumer markets being Germany (22 percent), the United States (15 percent), France (10 percent), the United Kingdom (10 percent), the Netherlands (9 percent), Japan (6 percent), Italy (5 percent), and Switzerland (5 percent).

The Netherlands remains the center of production for the European floral market, as well as a major international supplier to other continents. The flower auction at Aalsmeer is the largest flower market in the world. Since the mid-1970s, the production and distribution of cut flowers in Netherlands has burgeoned. In 1995, Dutch growers produced over 8 billion blooms and the flower auctions collectively traded more than 5.4 billion guilders (about $3.2 billion) in cut flowers and potted plants, contributing over 4 billion guilders annually to the Dutch balance of trade.

4.1.2 Production of Cut Flower across World

According to the 2009 Census of Horticultural Specialties production of cut flowers is worth over $400 million annually in the United States. More than 90 percent of that production was sold on the wholesale market to florists and other outlets. These data do not reflect the contributions of small-scale and part-time growers with sales less than $10,000. Many growers are producing a small amount of cut flowers to enhance their direct-marketing product mix. With few exceptions, most producers grow fewer than 5 acres of flowers.

At present, the vast majority of cut flowers are imported from overseas. Leading producers include the Netherlands, Columbia, Kenya, and Israel. Flowers imported from overseas are largely roses, carnations, Gerbera daisies, garden mums, and orchids. These flowers ship reasonably well and make up the bulk of the flowers used in arrangements by most florists. Most of the flowers that local growers focus on are those that do not ship well or have shorter postharvest vase lives. These flowers have come to be termed "specialty cut flowers." Examples of specialty flowers include sunflowers, zinnia, lisianthus, dahlia, ageratum, and peonies to name but a few on this long list. Local growers can readily develop a market niche with these flowers by using the advantage of longer vase life if produced locally, higher percentage of usable flowers, and a wider choice of colors and varieties.

Increasing demand for a wide variety of locally grown, fresh-cut flowers has kept this market growing in volume for years. Producing fresh-cut flowers is not for everyone, however, as they have special production requirements, as well as a fairly short shelf life. Any grower considering cut flower production should also be aware of the relatively short field growing and marketing season, especially those in the northern United States.

Fresh-cut flowers are generally sold either by the bunch, in prearranged bouquets, or individually. Some growers will set up u-design displays at markets along with pre-made bouquets. Six basic marketing alternatives are available to the cut-flower grower: wholesale markets, direct to florists, pick-your-own operations, roadside stands, farm markets, and subscription. Floral cooperatives, while standard in much of the rest of the world, have yet to take root here.

Because freshness is always a consideration, astute cut-flower growers should always seek to streamline the process from the farm to final consumer. As with marketing most annual crops, entering the market early with a quality product will help attract (and retain) customers to your business and usually command a higher selling price.

When planning production, first consider your ability to market. You should conduct some market research because growers often overestimate their ability to sell in a given market. Production of less than one acre of many floriculture crops is typical for most growers.

With the wholesale option, you typically either deliver your crop to a distributor or have them pick up the flowers at your farm. Wholesalers generally buy and resell your flowers for a predetermined price that can change during the season. This marketing alternative is subject to the greatest price fluctuation as product rapidly moves in and out of the wholesale marketplace. Wholesale prices also vary based on regional, national, and international flower inventories.

Florists should always be a primary consideration when developing your flower marketing plan. Market surveys indicate that many florists show at least some willingness to purchase locally grown cut flowers. With the generally high quality and wide selection available through florists' regular market channels, their expectations for service and quality are high. Emphasis on a broad selection and freshness may open florists' doors to locally produced cut flowers. Some growers start selling to florists by bringing a van-load of flowers to their shop and inviting the staff out to check out their selection.

Pick-your-own, roadside stands, and farm markets (both on the farm and at organized farmers markets) have the potential to generate substantially higher prices for the grower. Most new growers enter flower marketing through one of these avenues as they provide the greater level of inventory flexibility that new growers will likely need as their production skills develop. For more on establishing a roadside market, see Developing a Roadside Farm Market .

Subscription sales are an offshoot of the Community Supported Agriculture (CSA) movement. Most CSAs are based around produce sales with each member purchasing a share of a farm's production. This is a contract between the growers and their members that provides for a weekly portion of fruit and vegetables based on production and subscription level. If a crop is lost, everybody loses; if there is a bumper crop, everybody wins. Some CSAs provide flower shares as part of the contract or as an option. Another option under subscription sales is in providing vase service to restaurants, country clubs, and professional offices. Most flower prices are quoted for bunches of ten with specific stem lengths or individually for large flowers, such as sunflowers, or very expensive items, such as orchids.

Cut flower production has a long and distinguished history. Roses, carnations, and chrysanthemums have been cultivated for centuries in many parts of the world. These popular garden flowers are enshrined in literature and the arts. Gardeners would cut and bring these and other flowers into the home. In the 1600 and 1700's formal cut flower production became established in The Netherlands with the development of greenhouses. Spring flowers were brought into bloom early in the winter. In particular, lilac bushes were dug out of the field in the fall, subjected to natural cold temperatures to break dormancy and forced in greenhouses. The cut stems were then harvested and sold. Cut flower producers in the United States quickly followed European settlement, with the first greenhouses reported in the mid to late 1700's. Small producers were located near cities and towns to supply the populace with cut flowers. At that time transportation was slow and lacked refrigeration.

With the spread of Europeans across the continent, cut flower production moved with them. Production was not limited to greenhouses as there was also some harvest from outdoor grown crops. The development of air transport and refrigerated trucks greatly accelerated the concentration of the cut flower industry and allowed it to move to areas with the best climates for optimum production and lower production costs. Carnation and rose production moved to the Colorado mountains and then to coastal California. Chrysanthemum and gladiolus production moved to Florida and California. Northern growers stopped growing many of their traditional greenhouse crops. Field production at this time was declining, except with a few species such as gladiolus. California eventually dominated the fresh cut flower industry in the United States. Relocation of the cut flower industry did not stop with the move to California. Cut flowers could be imported from other countries because cut stems did not have roots or soil, which had restricted import of other floriculture crops such as plugs and potted plants. In the mid 1960's the first carnation production started in Bogota, Colombia. The industry rapidly increased because of the excellent climatic conditions – high light, moderate temperatures, and very low production costs, especially labor and heat. The production of chrysanthemums and roses followed in later years. Other countries followed suit, especially Ecuador, but none have been as successful as Colombia.

Today, the majority of all three species sold in the United States are imported. The rapid increase in imported flowers had a devastating impact on cut flower production in the United States as the majority of the carnation, chrysanthemum and rose producers shut down or converted to other crops, such as potted and bedding plants. A number of large producers still remain by focusing on a broad range of crops. As the cliché goes, "when one door closes, another opens". The loss of many large scale producers provided niches for small local producers to compete. They could provide high quality flowers which have never been boxed and shipped dry. They especially focused on flowers that do not ship well. At the same time during the 1990's retailers and consumers increasingly wanted more choices than just a few species. The classic "roundy moundy' rose, carnation, and chrysanthemum arrangements weren't enough for most consumers anymore. They wanted more interesting, less structured arrangements, mixed bouquets, and even vases filled with just one type of flower. They also wanted to see new and unusual flowers. Small producers resurged selling a broad range of cut flowers directly to consumers through farmers markets, roadside stands or even the occasional U-pick operation. Other growers targeted specialty florists, supermarkets and wholesalers. Much of the production is in the field, but growers also use an array of structures ranging from fully-controlled greenhouses to minimally heated hoop houses to unheated high tunnels.

4.1.2 Production of Cut flower in India

In the recent years in the rice producing belt of Asia crop diversification has become a popular strategy to maximize the use of land, water and other resources. If carried out appropriately, diversification can be used as a tool to augment farm income, generate employment, alleviate poverty and conserve precious soil and water resources. In the state of West Bengal in India, diversification in cropping pattern is a very recent phenomenon other than the case of plantation crop from the pre-independence period. It is basically a rice producing state which has achieved a very high level of productivity in paddy cultivation during the second phase of Green Revolution and is now the largest producer of rice in the country. However diversification towards high value crops is being considered as a way to increase the contribution of nonrice crops to output ratio to attain higher agricultural growth rates in the future. Moreover in a state where the fragmentation of land size has taken place to such extreme that the size of land holding of more than eighty percent of the farmers is less than 0.5 acres diversification in agriculture is now becoming almost a necessary feature for the small and marginal farmers as the traditional cultivation of paddy is becoming non-viable. While high value crops have been gaining in land area floriculture, also is increasingly being regarded as a lucrative opportunity to increase per unit return.

The art of growing flowers is not new to India as well as in the state, growing loose flowers mostly for worshipping, garlandmaking and decoration form the backbone of the Indian floriculture, which is mostly in the hands of small and marginal farmers. Twenty years ago, the growth was mainly focused on foliage plants for household purposes. The impetuses in cut flowers like rose, gladiolus, tuberose, chrysanthemum, carnation, gerbera, anthurium, orchids and lily has acquired momentum in the recent years. Enormous genetic diversity, varied agro climatic conditions and versatile human resources offer the state a unique scope for diversification into new avenues which have not been explored to a greater extent till now. If exploited in a proper manner this might help in augmenting the revenue of the impoverished farmers of the state.

Demand for flower can be categorized into two different types – one is the regular demand and other is the special demand. Regular demand again can be categorized in to two broad types – one is the original household demand for traditional flowers and the other is the demand for exotic flowers by the different hotels and offices which is the result of changing life style. This demand more or less remains steady throughout the year, though it has been on the rise over the last few years. Special demand on the other hand has increased immensely over a period of last five years or so. This type of demand is associated with special occasions like marriage ceremonies, birthday parties, Bengali New Year, all National Days, Valentine's Day, any cultural meet, Durga Puja and Other Pujas (religious festivals).

The study is based on a primary data survey carried out in collaboration with National Bank for Agricultural & Rural Development, (NABARD), India, in the major flower producing districts of West Bengal which are Nadia, North 24 Parganas, South 24 Parganas, Howrah, PurbaMedinipur, Hooghly, Darjeeling and Jalpaiguri. The flower producing areas were purposefully selected in consultation with the District Horticulture Mission Office. Some members of the local Flower Producers' Association and members of Gram Panchayats were instrumental for selection of the blocks and villages. The study was conducted in the state during January 2010 and June 2010. A two stage sampling technique was used to select the producers. A survey was also conducted in the four major market sites, namely Jagannath Ghat, Thakurnagar Bazar, Nokari Bazar and Dhantala Bazar to collect primary information on marketing and price trend of different variety of flower products.

Small and marginal farmers account for about three-fourth of the total operational holdings in the state of West Bengal. They operate over one-fourth of the total area cultivated and their average land holding varies between 0.5 ha to 1.2 ha. Majority of these small and marginal farmers cultivate mainly low value, subsistence crops of paddy. In the absence of adequate farm and non-farm employment opportunities, they are also forced to live below poverty line.

Since the time between planting and harvesting is usually three to four months in case of these flowers, therefore average cost has been calculated with three months time period. Any cost borne by the farmer after these initial three months is minimal and are therefore ignored. The following tables analyze the variable cost of producing cut flowers Rose, Gladiolus and Tuberose by the selected farmers. The price of the factors are recorded as reported by the farmers and are measure of actual cost incurred by them.. This calculation, however, does not account for overhead costs, though the opportunity costs of the fixed asset are small.

In case of flower cultivation the initial cost at the time of plantation is the highest and later on the costs become minimal. While the basic factors of production remains the same the pattern of utilization differs in case of each flower. The factor inputs are common to all the varieties. Also during survey it was discovered that as the land size differed widely the requirement of input quantities varied and so did the rates of some factor inputs. Unlike any other crops, cost of flower production therefore presents a tremendous challenge in terms of factor utilization analysis. The values of the explanatory variables like labor, non-human labor, land, fertilizers, pesticides, irrigation and agricultural machineries have been collected directly from the farmers of the different villages during the time of survey.

Gladiolus is one of the most important bulbous flowering crops grown commercially. It is grown abundantly in Darjeeling and Jalpaiguri districts by rotation almost throughout the year and also in the districts of North 24 Parganas, South 24 Parganas and Nadia in the winter months. In the blocks of Kalimpong I and Mirik of Darjeeling farmers are concentrating on gladiolus cultivation for a long period of time. The variety of gladiolus grown here are also many.

In growing gladiolus the major expenditure is in case of land preparation. The reason behind it is that in the climate of the districts of Jalpaiguri, North 24 Parganas, South 24 Parganas and Nadia are not always conducive for growing Gladiolus. The farmers instead try to combat this factor with better land preparation. In case of Darjeeling there is a unique situation pertaining to mountains. Since only about one third of the land can be cultivated of an acre, a farmer who is growing flower on one acre of land is actually cultivating in three acres. The land value is calculated accordingly. That is why the land component is the highest in case of Darjeeling. The next important component of cost here is the bulbs. Bigger and better quality bulbs are higher priced. Very little labour is required for growing gladiolus and farmers in most of the cases utilize family labour in this case.

The main sources of market information are traders and brokers. The majority of farmers become aware of the price upon their arrival at the market place. In most cases, the prevailing prices from the previous market days are considered when the farmers deliver the flowers to the wholesalers without allowing remunerations for the difference in quality, seasons and the change in market conditions. Brokers often deliver information about declining prices in the terminal markets to farmers so as to convince them to receive lower prices for what they will sell. Traders themselves lack market information outside their vicinity. Traders in Nokari Bazar and Thakurnagar Bazar, for instance, do not get precise information about prices in Kolkata markets to adjust the sales prices.

4.2 MARKET DEMAND FOR CUT FLOWERS

4.2.1 Introduction

When a grower decides to produce a certain product the decision is influenced firstly by the constraints of the production area and secondly by the market research conducted before production commences. Potential crops are not just limited to cut flowers but may also include dried flowers, native flowers, cut foliage and fillers for bouquets. Most growers increase their chance to make a profit by growing more then one variety throughout the growing season by choosing species that extend the harvest period. Successful growers will understand the limitations of the growing area through soil analysis, climate, aspect, drainage and irrigation and also the specific requirements of the varieties they choose to grow i.e. soil pH, fertiliser etc.

Cut Flower is termed as any flower that is cut from the plant, with its thorns trimmed, and after that flowers are ready to be used in a market for various end use. Cut flowers packaging is grouped into two categories such as single cut flower, and the bunch or a bouquet of cut flowers. Cut flower packaging includes carnations, roses, chrysanthemums, orchids, lilies, and other flowers. For cut flower packaging, flowers are cut with branches, stems, and leaves to be used for bouquets or decorations. There is wide range of packing containers for cut flowers, but most are of them are long and flat in shape. The most demanding product type of cut flower packaging is plastic sleeves to wrap flowers for any festivals, official and unofficial (social) events. Cut flower packaging enables individuals to make their ornamental products even more attractive and demanding. For retail market cut flower packaging is done to meet the requirements of both supply chain and retailers demand.

Producers close to their markets have a competitive edge i.e. The Netherlands and Germany. Smaller growers may decide to supply the domestic market only or may find a niche market for exports of specialty products i.e. native flowers. Flowers can be sold through wholesalers at markets to small local outlets, or at the farm gate. The small beginner may find it easier to start with local retail outlets, the local farmer's market and farm gate sales and even the internet, and then gradually branch out to larger distributors as production increases. Wholesalers usually require specific grading, packaging and a consistent quality, although prices will be lower then through direct sales wholesalers will handle large quantities for the grower.

Exporters have specific problems they need to overcome in order to satisfy potential export markets. Quality is probably the most important element, an efficient transport system is vital in retaining product quality as is the production system used and the harvest techniques including handling and post harvest handling.

The growth in the cut flower packaging market is primarily attributed to the evolving lifestyle of consumers, rising environmental concerns, the growth of ornamental packaging market and rising disposable income driving the growth of the cut flower packaging market. In addition, rising urbanization, particularly in the developing countries and increasing demand of sleeves cut flower packaging are the key factors responsible for the overall growth of cut flower packaging across the globe. Nowadays, cut flowers are increasingly used as a decorative material across the world. Furthermore, changing lifestyle preferences, increasing demand for flexible packaging, and growing production of flower across the world especially in Europe, North America and Asia pacific is anticipated to drive the cut flower packaging market. On the other hand, the major factor restraining the growth of cut flower packaging market is storage of the packed material and quality issues such as damage and spoilage of flower.

4.2.2 Cut Flower Packaging Market

The global cut flower packaging market is segmented into three parts based on the material type, product type, cut flower type, distribution channels, and geography.

Based on the material type the cut flower packaging market is segmented into:

- ☆ Cardboard
- ☆ Plastic
- ☆ Fabrics
- ☆ Jute

Based on the product type cut flower the global cut flower packaging market is segmented into:

- ☆ Sleeves
- ☆ Boxes
- ☆ Poles
- ☆ Bags
- ☆ Metal Stand

Based on the type of cut flower the cut flower packaging market is segmented into:

- ☆ Single cut flower
- ☆ Bunch or a bouquet of cut flowers

Based on the type of distribution channels the cut flower packaging market is segmented into:

- ☆ Direct Selling,
- ☆ E-commerce
- ☆ Online retailing
- ☆ Others
- ☆ Cut Flower Packaging Market: Regional Outlook

Regarding geography, the cut flower packaging market has been categorized into five key regions including North America, Latin America, Europe, Asia-Pacific and the Middle East & Africa. The cut flower packaging market is expected to register healthy CAGR during the forecast period. North America is experiencing the rapid increase in the production capacity of cut flower packaging market as a demand of floral industry is rising rapidly. The Europe cut flower packaging market is projected to register healthy growth as Netherlands accounted for around 52% of global cut flower market share and has been at the center of cut flower production in the European floral market. Also, the demand for the cut flower packaging is very high in US and Canada increasing production of flower, due to the changing lifestyle, and growing demand of the floral industry. Cut flower packaging market for Asia Pacific is also expected to witness steady growth during the forecast period primarily attributed to the growth of packaging market, especially in China and India.

4.2.3 Cut Flower Packaging Market: Key players

Some of the key player's global cut flower packaging market are T & G Flower Growers, Perlite Vermiculite Packaging Industries, Inc., Chrysal International, Dongguan Yason Pack Co., Ltd., Shenzhen CyGedin Printing & Packaging and many others. The capacity of cut flower packaging market to meet the increasing demand for the sleeves packaging by developing new production plants or by the product type advancement.

4.2.4 Demand of Cut Flowers in India

The floriculture industry faces many challenges including increasing energy and input prices, seasonality of its products and international competition. This study analyses the factors determining the supply and demand of fresh-cut flowers and uses simulation analysis to decompose it to analyze the production and productivity on a particular land area under cultivation. By understanding the factors which determine supply and demand, this study analyses the movement of the demand and supply curve. The simulation analysis performs a regression analysis to understand the productivity based on the past and current records so that appropriate measure can be taken to increase or decrease productivity in order to meet the supply and demand.

The floral industry is one of the major industries in many developing and underdeveloped countries. Floriculture as an industry began in the late 19th century in England, where flowers were grown on a large scale on the vast estates. The present day floral industry is a dynamic, global, fast-growing industry, which has achieved significant growth rates during the past few decades. Floriculture is in a heavy consolidation process driven by price pressure in retail which affects also the breeding companies.

As per the Department of Horticulture Tamil Nadu, most of the cut flowers are managed and processed by private nurseries. According to the states, there is 70% domestic market and 30% exports. On the other hand, Loose flowers, mostly owned by State Horticultural farms spanning across 700 hectares, cater to the 50% domestic market demand.

Cut flowers are parts of plants, characteristically including the blooms or "inflorescences" and some attached plant materials, but not including roots and soil. Fresh cut flowers are highly perishable because they maintain only limited life-supporting processes by taking water up through their stems. Fresh cut flowers are used for decorative purposes such as vase arrangements and bouquets at formal events; designs for weddings and funerals; gifts on occasions such as Mother's Day, Valentine's Day, in times of illness, and at holidays such as Christmas and Easter; corsages and boutonnieres; and informal displays to beautify homes and public places. Cut Flowers are available at the florist or can be cut from the home garden.

Preserved flowers are fresh cut flowers that have been dehydrated, preserved with a chemical solution and then air- or oven-dried. They may be used in boutonnieres, corsages, wreaths, formal and informal displays, and similar ornamental articles. Preserved flowers, known in the industry as "everlasting flowers" or "everlastings," are not as perishable as fresh cut flowers.

Factors affecting Demand

Income of People: Income is a major factor in determining the demand of any product. When the income of people increases, their tendency to buy increase, thereby increasing the demand of products. This increases the equilibrium price of the commodity. Similar is the case with the demand curve of flowers. During festival times, be it Diwali, Id or Christmas, due to the decorative value of a flower, they are always in high demand. Therefore the demand curves shifts to right and causes the equilibrium price to rise.

- ☆ Seasonality: Cut flower is a product which has an extremely high seasonal value. This means that a particular flower is highly demanded during a particular period of the year. This makes the movement in the demand curve very often. The responses to seasonal change depend upon the elasticity of that particular flower with respect to its decorative and religious value.

Price of Related Goods: The related goods in case of a product is determined by its substitute and compliment of it.

- ☆ Substitutes: Substitutes for fresh cut flowers are other objects which people use for decorative purpose and as a gift. Therefore a rise in the price of Greeting Cards, gift items will have an effect upon the demand for the flowers.

The cheaper prices or the fall in prices of the gift items would in turn cause a lesser demand for flowers. People would opt for gifts items other than cut-flowers to gift their loved ones. Such change of choices causes the demand of the flowers to change marginally.

- ☆ Compliments: Compliments for fresh-cut flowers are the horticulture industry which operates in tandem with the floriculture. An increase or decrease in the prices of items like seeds, manure, fertilizers, labor minimum wage have direct impact on the demand and thereby the equilibrium price. This also includes the fertility of the land in a particular state. An efficient productivity area of land under the production is major factor in determining the price of the fresh-cut flowers.

Demand of Interior Designers and Wedding Planners: With the onset of the demand of the Interior designers and Wedding Planners in contemporary times, the cut flowers' demand has also increased. Interior Designers and Wedding Planners make use of decorative flowers to enhance the beauty of their work. Therefore, if more and more people hire Wedding Planners and Interior Designers, chances of the demand of cut flowers to increase are high.

- ☆ Decorative Value of Fresh-Cut Flowers: Beautiful flowers, with respect to the common perception which people possess, are often in more demand. Flowers like Roses, Sun-Flowers, Carnation are always in high demand in times of festive times.

Factors affecting Supply

Seasonal Rainfall: The amount of rainfall immensely affects the production output of flowers, which indirectly affects the quantity of flowers supplied in the market. In a season when there are high amount of rainfall, the productivity escalates thereby shifting the supply curve.

- ☆ Establishment of Agriculture Economic Zone: 20 Cut Flower units are currently operating within this AEZ with focus on fresh cut roses, gerbera, carnation etc.
- ☆ AEZ for Flowers at Nilgiris: An AEZ for Flowers at a cost Rs.15.89 crores has been established at the Nilgiris. Facilities such as pack-houses, information centers, cold storage, model floriculture units are being set up.
- ☆ The Tamil Nadu has only one cold storage designated for fruits. This is located in Kancheepuram district and has a capacity of 2500 tons

Government Agriculture Reform: Government of India initiates numerous policies which affect farmers in both the ways. The amount of tax levied by the State government on the production of certain flower seed is high determinant factor in the pricing of flowers. Also the amount of rebate given to farmers on growing a particular also affects the supply chain scale cycle largely.

Awareness among farmers: Numerous awareness programs play a key role in increasing the production capacity of the farmers. Once the farmers are educated about the techniques which increase their production, will highly effect the supply in the market.

Foreign Trade Policies: The foreign trade policy in 2004-09 emphasized the need to boost agricultural exports, growth and promotion of exports of horticultural products. The tax levied by the government on the export of flowers determines the market price of the flowers in the domestic market.

Public as well as Private Investments: State as well as Corporate Investments are required in modern methods of grading, post-harvest management and development of cold chains. Such investment brings in new players, including large corporate players to set up new manufacturing units in the country.

Shipping Capability of the Breeder/Producers : The amount of supply in the market of the flowers depend upon the capability of the Breeder to ship/transport the amount of flowers demanded for a particular purpose. The charges put up by the transport companies also add up to the cost of the flowers in the retail market.

4.3 PRODUCTION REQUIREMENT OF CUT FLOWER

4.3.1 Site Selection

The vast majority of cut flowers prefer a well-drained site. Having deep, fertile soils will greatly simplify the process of site development, but soils can be improved through cover cropping, the application of composts and manures, and other soil-building practices. Conducting a soil test prior to planting and applying lime and fertilizer according to the soil test recommendations is highly recommended. Soil test kits may be obtained at your local extension office. Installing drain tile or building raised beds may offer an adequate solution for sites with poor drainage or very shallow soils. Raised beds are fast becoming a standard practice in growing higher-value produce and flowers. They provide for an earlier crop by allowing the beds to be heated by the sun on the sides as well as on the top. They can also boost production by increasing the depth of topsoil. When used with plastic covering for weed control and drip tape for fertigation, raised beds can greatly decrease weed pressure and simplify fertilizer application.

4.3.2 Planting

Most new growers begin by planting everything by hand. The need to plant many different flowers in small plots over a long time period tends to keep planting largely a hand operation. Plants such as sunflowers are both direct seeded and transplanted, while others are best started as plugs (starting the plant in a greenhouse or cold frame in flats with small cells in a planting medium) and then transplanted to the field or bed. While traditional vegetable transplanters can easily do double duty in planting many types of cut flowers, many growers continue to place all their plugs by hand. Spacing between plants varies widely within each species and sometimes by individual cultivar as some plants are much larger than others. In general, cut flowers are planted closely together to encourage growth of longer stems. New growers will want to learn the specific demands of each flower in order to get the best production.

4.3.3 Irrigation

Any high-value crop should be irrigated to ensure more reliable production. Most flower growers use drip irrigation tape either under plastic or laid on top of the ground to maximize water efficiency, keep water off the leaves and flowers (this improves overall quality as it prevents foliar diseases and chemical stains), and provide for fertilizer or chemical application. In addition, a steady, reliable supply of water ensures production of the longest stems possible.

4.3.4 Wind Protection

Producing high-quality cut flowers requires preventing wind damage since bent flower stems have little or no value. Therefore, wind protection through both site selection and use of windbreaks is necessary when planning your operation. Windbreaks can be live plants, such as an evergreen hedge, or fencing materials, such as split bamboo or board-on-board planking. Whatever method you choose, it is important to consider how far the most distant flowers are from the windbreak. The farther away the windbreak is, the less effective the protection. Sometimes, multiple windbreaks are required. Many cut flowers require stem support as well. Several wide-mesh products are available that can be used as a grid for the flowers to grow through, thus giving them support. Support mesh also prevents damage by keeping flower stems upright during heavy rain.

4.3.5 High Tunnels and Greenhouses

The short outdoor production season in the northern United States has encouraged the adoption of growing structures to lengthen the marketing season. Having overhead cover also prevents damage to blossoms due to rain. Unheated greenhouse frames, such as high tunnels, can meet this need with a modest investment. These structures provide ventilation by rolling up the sides and, if equipped, through roof or ridge vents. The additional radiant heat stored in a high tunnel allows for a planting date 4-6 weeks prior to the outdoor season and may allow another 4-6 weeks of production in the fall after the first frost. Some growers place small heaters in their tunnels to allow for an additional 2 or more weeks of season extension.

A greenhouse is a climate-controlled environment with heat and ventilation carefully managed for maximum plant growth. Greenhouses can look very similar to high tunnels, but they are much more expensive to construct and maintain due to their more sophisticated structure and environmental controls. Producing flowers in a cost effective manner may not be possible in a greenhouse due to the high cost per square foot of production area. Many growers use smaller greenhouses to produce transplants for use outdoors or in high tunnels versus using the space for flower production.

4.3.6 Pest Control

Cut flowers represent a wide group of plants. There is a significant difference in pest problems among each genus, species, and cultivar. Discussion of the range of pest-management options available to cut-flower growers is beyond the scope of this publication. As a new grower, you will want to do some careful research into each plant that you choose for your operation. When possible, select those that are indicated as relatively pest free. Keep careful records of those that you do select and scout your plants often for pest problems. Detecting a pest problem early will usually make controlling it easier and less costly.

4.3.7 Insects

While a wide variety of insects will inhabit any flower planting, only a few are recognized as causing economic damage, including aphids and thrips. There are wide differences in not only pests but also tolerance to specific pests. Biological, biorational, and synthetic chemical controls are available. Pest populations can also be controlled by naturally occurring parasites and predators. Avoiding cover sprays of broad-spectrum insecticides will help maintain these populations. Beneficial insects, such as lacewings, can also be purchased from biological control companies. Biorational controls include horticultural oils, insecticidal soaps, and plant extracts such as neem oil. These typically have short residuals and minimal impact on beneficial organisms. It may be necessary to use synthetic chemicals for insect control at times. Always use products labeled specifically for the use you intend and seek to use materials that have the lowest toxicity to bees.

4.3.8 Diseases

Fungal diseases can have a major impact on cut-flower plantings. While many diseases will not kill the flowers, they can drastically reduce their marketability. Powdery mildew can be a real problem in zinnias and many other flowers. Many seedlings are susceptible to damping off, and a number of leaf spot fungi can also reduce the value of certain species. Good crop rotations, careful selection of cultivars, maintenance of optimal growing conditions, and control of weeds and insects will significantly reduce the incidence of diseases.

4.3.9 Weeds

Weeds are generally controlled best through good preplant soil preparation. Using a cover crop will not only add valuable organic matter to increase soil tilth, but many cover crops will also outcompete or suppress weeds. Cereal rye, for example, can provide significant weed control through smothering and suppression from natural herbicides produced by the rye. Other cover crops, such as hairy vetch, will add significant nitrogen to the soil and attract many beneficial insects as well. These cover crops are traditionally planted the prior year and are usually tilled under prior to planting your flower crop.

Using a preemergence herbicide labeled for use in flowers is another avenue worth considering. Only a handful of herbicides are currently labeled for flower production. In general, it is very easy to find an herbicide to control both pre- and postemergent grasses in flowers, but the choices for broadleaf weed control are very limited because many flowers are closely related to these weed types. Planting plugs alleviates much of this problem as well-chosen preemergent herbicides seldom damage well-established plants. The Penn State Extension publication Controlling Weeds in Nursery and Landscape Plantings (available for free through your local extension office) is a good source of information on herbicide use in flower production. Make sure you read the product label and understand the proper uses of any herbicides before you apply them.

Many growers decide to use plastic mulch for weed control. This has many advantages over herbicides in that it works on all weeds and will prevent mud from splashing onto the flowers as well. Until recently, the only choice was black plastic, which worked well. Now research is exploring many of the new, colored plastic mulches. In vegetable production, research conducted on some of these colored mulches has shown them to increase yield and decrease pest populations.

Some of the pesticides that may be used in cut-flower production are restricted-use pesticides and require a pesticide license to purchase. Pesticide applicators tests are usually administered at county extension offices, so you should contact your local office for dates and times of these examinations. When using any pesticides in your enterprise, remember to follow all label recommendations regarding application rates and personal protection equipment (PPE) requirements. Also remember that any Worker Protection Standards (WPS) apply both to you and your employees.

4.3.10 Harvest and Postharvest Handling

Flowers are best harvested in the morning after the dew has dried but before the heat of the day because the flowers are cooler and will suffer less heat stress and better retain their quality. Packing damp flowers close together for shipping will promote the development of Botrytis (gray mold), which will render the flowers unfit for sale. Rapidly cool the harvested flowers or at least remove them from direct sunlight. Many flowers benefit from the use of floral preservatives. Some flowers, such as yarrows, are best harvested as soon as pollen is evident, while sunflowers are best harvested as soon as the flower is almost completely open. Again, knowing the characteristics of each flower you decide to grow is critical to producing a high-quality product. Planting small test plots prior to entering actual production will aid in the flower-selection decision-making process.

4.3.11 Environmental Impacts

In the normal course of operations, farmers handle pesticides and other chemicals, may have manure to collect and spread, and use equipment to prepare fields and harvest. Any of these routine on-farm activities can be a potential source of surface water or groundwater pollution. Because of this possibility, you must understand the regulations to follow concerning the proper handling and application of chemicals and the disposal and transport of waste. Depending on the watershed where your farm is located, there may be additional environmental regulations regarding erosion control, pesticide leaching, and nutrient runoff.

4.3.12 Risk Management

First, you should insure your facilities and equipment. This may be accomplished by consulting your insurance agent or broker. It is especially important to have adequate levels of property, vehicle, and liability insurance. You will also need workers' compensation insurance if you have any employees. You may also want to consider your needs for life and health insurance and if you need coverage for business interruption or employee dishonesty.

Second, check to see if there are multi-peril crop insurance programs available for your agricultural enterprises. There are crop insurance programs designed to help farmers manage both yield risk and revenue shortfalls. However, individual crop insurance coverage is not available for all crops. If individual coverage is not available for what you grow, you may be able to use the AGR/AGR-Lite program to insure the revenue of your entire farm operation. To use AGR-Lite you must have five years of Internal Revenue Service (IRS) Schedule F forms. For more information concerning crop insurance, contact a crop insurance agent or check the Penn State Extension website.

Finally, the USDA Farm Service Agency has a program called the Non-Insured Assistance Program (NAP) that is designed to provide a minimal level of yield-risk protection for producers of commercial agricultural products that don't have multi-peril crop insurance coverage. NAP is designed to reduce financial losses when natural disasters cause catastrophic reduction in production. NAP coverage is available through your local USDA Farm Service Agency office. The application fee for this program may be waived for eligible limited-resource farmers.

4.4 SOILS AND NUTRITION

The major goals for an outdoor cut flower fertilizer program are to provide adequate levels of nutrients for vegetative growth early in development, promote the development of strong stems and encourage a consistent yield of good flowers throughout the cutting season. In addition to producing a good crop, another important goal is to use fertilizers in a responsible manner to maintain soil health and avoid environmental contamination.

Generally, nitrogen fertilizer for cut flowers should be applied on the basis of 1-2 lbs. of actual nitrogen per 1,000 sq. ft. Some large and fast-growing annuals like sunflowers may need more, and a slightly higher rate may be needed on sandy soils or where irrigation or rainfall rates are high. Generally, nitrogen fertilization involves split applications or applications targeted at the main growth period.

Adequate nitrogen is critical for cut flower production. A nitrogen deficiency will result in poor plant growth, a reduction in flower yield and the appearance of foliar chlorosis. Too much nitrogen may result into much vegetative growth, cause weak stems and delay flowering.

Unfortunately, nitrogen is required in the greatest amount by plants and is easily lost from the soil by leaching or in the form of a gas. For these reasons, nitrogen fertilizer must be applied each year to maintain adequate levels for production. Leaching of nitrogen as nitrate can be a serious problem not only as a way of losing nitrogen for the crop, but also because it can pollute groundwater and surface bodies of water. Too much nitrogen fertilizer, high rainfall or excess irrigation, and sandy soil are all factors favoring nitrate leaching. Nitrogen fertilization also affects soil pH. Over time, the continuous use of ammonium or urea fertilizers tends to lower pH, and lime stone may be required to neutralize the acidity.

4.4.1 Fertilization for Annuals

Germinating seeds and young seedlings are very sensitive tofertilizer injury, so initial fertilizer applications should be very small. Starter fertilizer should be applied at a rate of about 1/4-1/2 lb. of actual nitrogen per 1,000 sq. ft. at seeding either by broadcasting or banding. Bands should be placed about two inches below and to the sides of the seeds.

For annual species started from seed in the field, fertilizer should be applied after the young plants are well-established, according to the following recommendations for annual transplants.

When transplants from the greenhouse are used to start the crop, about 20-30 percent of the total nitrogen requirement should be applied in a preplant application made just before planting, assuming the transplants are healthy and have been hardened-off. Another 20-30 percent of the total should be applied as a side dress application when the transplants are about8-10 inches tall.

Sometimes, one application of fertilizer at planting can grow a good crop of annual cuts. However, split applications of nitrogen fertilizer are more efficient in fields receiving heavy irrigation or rainfall or that have sandy, well-drained soil. Split application provides a more consistent supply of nutrients than one large application made at planting and comes closer to meeting the actual needs of annuals as they develop.

4.4.2 Fertilization for Perennials

After flowering or harvest, most perennials continue to grow, building the root system and expanding the crown or other overwintering structures. So some perennials may benefit from a second fertilizer application after flowering. About 1-2 lbs. of actual nitrogen per 1,000 sq. ft. is required by peony. One-half should be applied at emergence, one-fourth just after flowering and one-fourth at dormancy. Asiatic and Oriental lilies grown as perennials should be fertilized by side dressing with 2-3 lbs. of 10 percent nitrogen fertilizer per 100 ft. row as the shoots begin to emerge from the soil and again at the same rate right after flowering. Late-season fertilizer applications to perennials are somewhat controversial. Some feel that fertilizing in the fall may disrupt the plants' normal preparations for dormancy and make them susceptible to cold injury, while others think there isa benefit to the roots and crowns. I recommend using caution and avoiding high nitrogen applications; however, applying phosphorus may be beneficial to root and crown development. Also, it would be safe to apply limestone if it is needed to adjust pH.

Just like in the greenhouse, complete nitrogen/phosphorus/potassium (NPK) fertilizers are applied to outdoor cut flowers according to the nitrogen requirement.

Ideally, however, phosphorus and potassium should only be applied if their need is indicated by a soil test. So rather than using 10-10-10, a grower could use single element carriers such as ammonium nitrate, pre-plant superphosphate and potassium chloride to fulfill the exact needs according to a soil test.

The effects of phosphorus on cut flower production are less obvious than those of nitrogen. A phosphorus deficiency may result in smaller plants and shorter flowering stems, but foliar symptoms may not be apparent.

Less phosphorus than nitrogen is needed by plants, and phosphorus is more stable in the soil than nitrogen. Some agricultural soils test "very high" or "excess" in phosphorus because of the regular use of NPK fertilizers over many years. Generally, the potential harm caused by high phosphorus is not to the plants but to the environment. Too much phosphorus can lead to runoff to surface bodies of water. This encourages excess growth of algae and other aquatic plants, which leads to a serious decline in water quality. Ideally, nutrients should be applied based on need as determined by a soil test. Unlike nitrogen, an application of phosphorus may not be needed every year.

Superphosphate is a soluble phosphorus fertilizer and is best used to correct a low-phosphorus condition before planting. Rock phosphate is very insoluble and is best used for the long-term maintenance of phosphorus in the soil. Potassium deficiency causes marginal chlorosis and burning on the lower leaves first. Unless the deficiency is severe, it may not affect the leaves on the Á harvested stems, but it could reduce overall yield. Potassium deficiency is most likely to occur on sandy soils low in clay and organic matter.

It may not be necessary to apply potassium fertilizer to some clay soils and soils high in organic matter every year. There is no harm to the environment from excess potassium and no direct harm of high potassium to most plants. However, too much potassium can depress the uptake of calcium and magnesium, sometimes to the point that deficiencies of these elements develop.

4.5 PRODUCTION TECHNOLOGY OF ROSE, GERBERA, CARNATION, LILIUM

4.5.1 Rose

Rose is one of the popular, oldest, modern and lovely flower grown all over the world. Rose plants can be a kind of shrubs, climbing or trailing with stems that are often armed with sharp prickles. Flowers vary in size and shape and are usually large and showy, in colours ranging from white through yellows, reds or multi-colours. Most of rose species are native to Asia, with smaller numbers native to Europe, North America, and northwest Africa. Rose plants range in size from compact, miniature roses, to climbers that can reach 7 meter in height. There are many rose varieties developed in different regions of Asia. The rose plant belongs to the family of "Rosaceae" and genus of "Rosa". Rose cultivation is always in high demand due to national and international markets.There are many verities of rose flowers are available. However, before going for commercial rose farming, it is important to select the right cultivar for your region. Roses can be grown in pots, on terrace, indoor, open fields, green house and poly house. Greenhouse rose farming is becoming very popular due to controlled and protected environment. Greenhouse rose cultivation also produce quality of flowers with high yield when compared to open field cultivation.

Roses are adaptable to a wider range of climatic conditions and found widely in Asia. Rose plants can be shrub, climbers or tremblers with or without thorns depending upon the variety. Botanically, roses belongs to Rosaceae family and 'Rosa' genus. Cultivation of roses can be done at home level in pots, on backyard, fields, terraces or indoors. Commercial production of roses can be done both in open air and polyhouse but high quality roses like dutch rose is mainly done in polyhouse farming where the environmental conditions are under control. Higher quality roses with more yield are obtained in the green house. Commercial cultivation of roses can be highly profitable as there is a growing demand for the rose flowers as cut flowers and loose flowers in floral arrangements, making bouquets, gifting as well as for manufacturing rose based products like rose water, gulkand, perfumes and cosmetics.

There are 120 species of roses, distributed all over the world. So far as garden roses are considered only species are important. Rosa damascena, Rosa foetida, Rosa chinensis, Rosa gallica are some of the important species used for rose cultivation.

- ☆ Miniature Roses: It is also known as baby roses or fairy roses. It is mainly used for making bouquets and in decorations. It is also suitable for pot culture and for beautifying terraces and balconies. Some of the important miniature rose varieties are Pixie, Baby goldstar. Baby masquerade.
- ☆ Floribundas: It is also known as hybrid polyanthus. These are climbers which grows upright and needs support to grow.
- ☆ Bourbon Roses: Bourbon roses are also known as reunion roses.
- ☆ China Roses: It is known to be the ancestor of the present day popular roses. It has been considered as a refuge for decorative as opposed to exhibition roses.
- ☆ Polyanthus: Polyanthus roses are normally dwarf, bears immense clusters of small blossoms. Echo, Chatillon rose are some of the popular polyanthus roses.

- ☆ Multiflora Ramblers: It belongs to the ramblers group. It is suitable for covering fences, walls and trellis.

Roses in Rose Cultivation

Roses can be grouped into 3 classifications according to their growth characteristics.

Bush Roses: Basically, Bush roses are self supporting & bear flowers at the top of the plant. Plant height may reach up to 6 feet. Hybrid teas, Hybrid perpetuals, Floribunda roses, Grandiflora roses, Polyantha roses, Miniature roses and Tree or standard roses belong to this category.

Shrub Roses: These roses belong to a non-specific class of wild species, hybrids, and cultivars that develop large, dense growth needing little maintenance and they are hard in nature. Species and old-fashioned roses generally bloom only once per season. Old-fashioned roses were popular in colonial gardens for their fragrance. However, we cannot compare these roses with today's roses.

Climbing Roses: These rose plants are extremely vigorous plants with long branches that require support or training. These branches can be trained to a trellis or fence or allowed to sprawl as a bank cover. Canes may range in size from 5 feet to 20 plus feet depending on the type of rose cultivar and how they are supported. Ever blooming climbers, Rambler roses, Large-flowered climbers and Trailing roses belongs to this category.

Factors responsible for Rose Cultivation

- ☆ Soil and Climate Requirement: Rose plants require well-drained sandy loam soils rich in organic matter with pH value of 6 to 7.5 for best yield. These plants require minimum of 5 to 6 hours sunshine throughout the growth period. The temperature range of 15 °C (Nightly temperature) to 28 °C (Day time temperature) is ideal for rose cultivation. As these plants are sensitive to water-logging, make sure to have proper soil drainage to drain out the water in case of excessive water or flooding. Sunlight is very important during the flowering stage, in areas where mostly cloudy or moist conditions impact the yield of roses.
- ☆ Land Preparation:- The planting site (Bed/pits) should be prepared at least 4 to 6 weeks before planting in the main field. Give couple of ploughing and hoeings to bring the soil to fine tilth and weed free. Prepare the beds 60 to 90 cm deep. Refill the pits with the mixture of well decomposed farm yard manure (FMY) and top-soil. Irrigate the pits to settle down the loose soil. Adding more mixture of soil and manure for leveling the bed is recommended.
- ☆ Planting Season:- The best season to plant rose plants are just before monsoon.

- ☆ Propagation:- The rose crop can be propagated by rooted cuttings or by budding on Briar root stocks. One year old budded plants are recommended to plant in July to August at 75 x 75 cm distance.
- ☆ Planting and Spacing:- Planting should be done immediately when the new plants arrive from the nursery. Rose planting late in the afternoon is considered to be the best. There should be a distance of 75 cm between plants. Make a hole of about 30cm deep and 30 cm diameter in the bed; treat the soil taken out from hole with 5% Aldrin dust (about 30 gm/hole). Rose plants with earth-ball are then set into hole with main stem in center and at level as it stood in nursery. Hold the bush firmly upright with one hand and then return treated soil to the hole around the roots and firm with hands and thorough watering immediately. If there is strong sunshine, the plants should be shaded for 4 to 5 days and leaves sprinkled with water.
- ☆ Manures and Fertilizers:- It is better to make soil rich by supplementing with organic matter while preparing the land.. At 3 months interval, apply well decomposed farm yard manure (FYM) @ 10 kg and 8:8:16 grams of N:P:K/ plant after each pruning activity.
- ☆ Irrigation:- Irrigation should be carried out immediately after transplanting in the main field. The new rose plants should be irrigated frequently (once in 2 days) till they establish and settle. Subsequent watering should be done once a week. Irrigating the rose pants with salt water should be avoided. In case of heavy rains or floods, make sure the soil drains out the water quickly as these plants are very sensitive to water stagnation.
- ☆ Intercultural operations:- The following should be carried out as part of intercultural operations.
- ☆ Weeding: Weeding and hoeing should be carried out with an agriculture tool after every alternate irrigation; Usually, the suckers which are different in shape and colour appear near the base on the stem or come from the roots underground; these should be removed as soon as they appear on the pant. Flower buds or flowers from the new plants should be pinched during initial stages so that they attain good growth. Remove faded, dead or diseased flowers regularly.
- ☆ Pruning: All roses require pruning in the second year of their planting and subsequent years. Basic principles of pruning are common to all type of roses and the following are the pruning methods in rose cultivation.
- ☆ Pests and Diseases:- The following are the pests and diseases found in rose farming.

- Red-scale:Lives under reddish brown incrustation on the branches and sucks the juice of plant. Spray with 2 ml Folitheon or Cytheon or Metasystox in 1 litter of water once in April and again in September controls the infestation.
- Rose Chaffer: usually appear in July, August and eat away the leaves. Spray of one ml of Malathion or rogor in one liter of water is effective control.
- White Ants: It is recommended to treat the soil with a mixture of 10% of BHC and 10 % of DDT at the rate of 30 grams/ pit. One ml Aldrine in one liter of water may also be applied around the plant.
- Digger Wasp: Burrow into the stem pith through the cut ends immediately after pruning. Tamic is applied in the soil for control. Spray with 2 grams of sulmite-75 in one liter of water.
- Die-back: Spray with fungicide (0.2 Captan or copper oxychloride) soon after pruning.
- Powdery mildew: Spray with 2 grams of Karathane or Benlate or Bavistin in 1 liter of water.
- Black Spot: spray with 2 grams of farbam or any other fungicide in 1 lt. water. Diseased leaves should be collected and burnt.

☆ Flowering and Harvesting:- The rose plants will be ready for flowering in the first year and will give economic yield from the second year onwards. Harvesting is done with sharp secateure at the tight bud stage when the colour is fully developed and the petals have not yet started unfolding. Flowering will commence from 45 to 50 days after pruning. Fully opened flowers are picked early in the morning.

☆ Yield:- Yield depends on the cultivar (variety), soil, climatic conditions and orchard management practices. In rose cultivation on average of about 10 lakh flowers/ha/year can be obtained.

☆ Post harvesting:- Roses must be placed in a bucket of water inside the poly house immediately after harvesting and transported to cold storage (2 to 5°C). The length of time depends upon the variety and quality of the roses. The flowers should be graded according to the length. It varies from 40 to 75 cm depending on the variety and packed in 8/10 per bunch.

4.5.2 Gerbera

Introduction

Gerbera is one of the most popular flowering plants. Some say it practically sells itself! However, if quality is not there, sales will suffer. For the highest-quality finished product, Tom Linwick, grower account manager for Syngenta Flowers/Goldsmith Seeds, says gerbera production requires more attention to detail than most crops.

Gerbera is a profitable plant if it is done on a large scale considering good quality planting material, irrigation system, greenhouse facility and proper management of fertigation, pest and diseases. It is 24 to 30 weeks crop and the first flower produced after the 7-8 weeks after plantation when plants are with 14 – 16 leaves. Ongoing basis daily time required per 4000 sqm greenhouse is 4 hrs depending on the number of manpower.

Gerbera crop is sensitive to frost conditions so it is cultivated in subtropical temperate climate conditions under poly house. The production of quality flowers requires optimal day temperature of 20oC to 25oC and a night temperature of 12oC to 15oC.

In the early stages, gerberas require special attention to moisture management. In producing young plants, too much water between the second and fifth weeks results in plants that do not develop normally. Signs of overwatering are deformed leaves with a thick, leathery appearance. These deformed plants will not develop properly unless soil moisture is decreased.

Germination

Temperature: The optimal temperature is 74° to 75°F (23°C) for the first three to four days. Lower it after germination is complete to 70° to 72°F (21° to 22°C) for the next three to four days. Eventually lower the temperature to 68°F (20°C) for growing on.

Moisture: Keep them saturated the first three days or until radical emergence (approximately the fourth day). If you are unable to dehumidify on day 4, decrease the starting moisture level from saturated to a wet. Media moisture should alternate between a saturated to a moist within 18 hours at radical emergence. Around the fifth day, begin alternating between a wet and a moist level. Keeping the newly germinated seedlings on the dry side will initiate rooting and help prevent foliar disease. When cotyledons are fully expanded, begin to alternate between a wet and a medium moisture level.

Humidity: Grow at 100 percent humidity for the first three to four days. Reduce it to 75 percent for the next four to five days. After one week, avoid disease problems by lowering the humidity to 40 to 60 percent. A maximum of 70 percent humidity should be maintained. Good air circulation is very important.

Light: Optimum day length during germination is 16 hours. During periods of low light, use supplemental lighting at 35 to 40 watt m2 (3500 to 4000 lux), or 400 foot-candles. Gerbera require light during germination so the seed does not need to be covered. If you use vermiculite to maintain a higher humidity, use a very light covering so the seed is still visible after watering the trays.

Media: A porous, well-drained media should be used with an EC of 1.2 to 1.5 and a pH of 5.5 to 5.8. Use a coarse peat to provide better aeration.

Plug Crop Time: In a larger plug, the crop time is approximately seven weeks from sowing if proper temperature, light and fertility guidelines are followed. The young plants are ready to be potted when they reach the four to five true leaf stage.

Fertilizer: Begin feeding with a complete fertilizer after the first week or once the seedlings are established. Some fertilizers that work well are a 14-4-14 or a 17-5-17 at 50 ppm or even a 20-10-20 at 60 ppm if the light levels are higher. Make sure the plants receive enough feed and the pH remains between 5.5 and 5.8. After 14 days, increase the fertilizer level to 75 ppm.

Transplanting

Transplant on time so plants don't become root bound and delay flowering, inhibit development and result in plants that don't size up properly. Make sure they aren't planted too deeply. If the soil covers the crown of the plant, it will inhibit proper growth.

Growing Media

Use a fully fertilized, porous, well-drained soil. EC should be 1.5-2.0 with a pH of 5.5-5.8. A coarse peat with 20 percent perlite provides good aeration.

Temperature

Start at 70-72°F for two weeks and then lower to 65-68°F. Use a 5-10 degree negative DIF or drop in temperature two hours before sunrise until 9 a.m. You can also lower the temperature at sunset for the same result.

Moisture

Alternate between moisture levels of wet and medium. Allow media to approach medium before resaturating to wet. Do not allow plants to wilt, which can cause root injury.

Lighting

Supplemental lighting is beneficial during winter months. In the finishing stages, avoid a day length greater than 16 hours. When plants are getting established after transplanting, a 14-hour day length works well, and then after seven to 10 days, reduce it to 11-13 hours. The light intensity with supplemental lighting should be 400-600 footcandles (4,000-6,000 lux); 14-22 moles.

Fertilizer

Constant feeding with a 17-5-17 fertilizer at 150-200 ppm is needed to maintain an EC in the soil of 1.75 to 2.0. Under high light conditions, a 20-10-20 fertilizer can be used. Supply 8-12 ppm phosphorous and 1.5 ppm iron. The use of a calcium-based fertilizer is recommended during the winter months but may need to be supplemented with an ammonium-type fertilizer.

Watch your pH closely. The optimal pH is between 5.5-5.8. Gerberas favor a lower pH and will show signs of iron and manganese deficiencies when the pH reaches 6.5. At this point, the leaves will have a mottle to the foliage.

Growth Regulators

For smaller containers like 4-inch pots, apply B-Nine (daminozide) at 2,500 ppm seven to 10 days after potting. Do not apply B-Nine once buds are visible, because this will cause shortened flower stems. When pea-sized buds are visible, the plants will flower in approximately three weeks.

Diseases

Preventive fungicides may be applied for Rhizoctonia, Pythium and Phytophthora.

4.5.3 Carnation

Carnations are most important cut flowers of world. Most preferred flower for commercial purpose because of their excellent keeping quality, withstand in long distance transportation and remarkable quality to rehydrate. Europe and USA cultivated this crop on large area where as in India it is cultivated on smaller scale. Himachal Pradesh, Punjab, West Bengal, Jammu and Kashmir and Karnataka are major carnation producing states. Flowers come in different shades such as yellow, pink, lavender etc.

It can be grown on any type of soil but soil with good drainage capacity gives best results. Rich sandy loam or loamy sand is most ideal soil for carnation cultivation. pH of soil for optimum growth of crop is about 5.5 to 6.5.

Varieties

Flowers are group into two major classes:

- ☆ Standard type
- ☆ Spray type.
- ☆ Mignon and Micro type
- ☆ Dianthini type

Varieties in standard carnations are Ariane, Corso, Candy Master and Tanga, Solar, Star, Athena, Happy Golem, White Liberty, Emotion, White Dona, Lisa, Domingo,

Master, Gaudina

Varieties in spray carnations include Rhodos, Alliance, Barbara, and West Moon. Estimade, Indira, Vera, Durago, Berry, Orbit Plus, Sunshine, Autumn, Rosa Bebe, Spur, Suprema.

Varieties in mignon and micro show shorter stems and higher production as compared to spray type. It is a small group of varieties.

Varieties in dianthini type shows similarity to spray type, but this type of flowers are smaller and produce more flowers on long stems. Market of these types of carnation is small.

Climatic requirements

The ideal climate for carnation production should have a cool but stable temperature, low humidity and long days with high light intensity. The standard carnation prefers a cooler climate than the spray varieties.

Light

Carnations are long day plants and require high light levels to produce high quality flowers. About 21.5 klx is considered to be the minimum natural light intensity required for adequate photosynthesis of carnations.

Temperature

Temperature influences many factors of production and quality. Temperature at night is very important for the quality. The difference between day and night temperature should be minimum. The combination of low night temperature (13.3oC) and high day temperature (23.80 oC) is ideal for the growth. As reported, carnation cultivation temperature range 10-15oC is preferred in green houses.

Soil

Carnations can be successfully grown in any type of soil but should be well drained and be in good physical conditions. The soil is usually sterilized with chemical disinfection methyl bromide (30-60 gm per square meter). Steam sterilization or sterilization with 10% formaldehyde, solarisation are the other methods that can be followed for sterilization of soil. The soil pH of 6.5 is the optimum. E.C is 1.2 at the start and 1.5 at the generative period is ideal.

Bed layout

Normally for working convenience, a bed of one meter wide is created with the length being equal to length of polyhouse. Between the beds a working path of 45-60 cm can be given. Raised beds are ideal for cultivation. Bed can be ideally prepared by mixing of well rotten coir pith, sufficient amount of compost and sand depending upon the clay content of soil.

Propagation

Carnations are multiplied vegetatively through cuttings 10-15 cm cuttings with 3-4 nodes weighing around 10 g is ideal for multiplication. Rooting hormone such as IBA at 500 ppm is used prior to planting of cuttings for rooting. Terminal cuttings give rise to good plants. Cuttings can be stored at 0oC before planting for several weeks. Cuttings are spaced at 5 cm apart and intermittent misting should be used for good rooting. Cuttings normally develop good root system within 21days. The rooting medium should be sterilized before planting. Drenching with fungicide is ideal to control fungal problem during rooting.

Spacing and planting

Carnation plants are planted in different spacing. Normally, 30-45 plants per m2 is considered to be ideal. 15x8 cm, 15x15 cm, 15x20cm, 15x10 cm, are different spacing followed. Alternate planting is beneficial in terms of reduced incidence of disease. It is important to maintain the plant number. In carnation, shallow planting is followed. Shading should be given the humidity to prevent plants from drying.

Support material

Carnation crop has the tendency to bend unless supported properly. Good support material is metallic wire woven with nylon mesh. At every two meters the wire should be supported with poles. The poles at both ends of beds should be strong. Metallic wire is tied around the bed along the length with the support from supporting poles. Bottom net can be of 10x10 cm then two nets of 12.5x12.5 cm and the upper most can be 15x15 cm.

Pinching

The first pinch is done at 6th pair of leaf stage and 2 nd pinch after 7th week of planting. Depending upon the need of crop spread, Single, One and half and Double pinches are given. Single pinch given below 6th node, to get early crop. Each shoot should then developed pairs of leaves about 3 weeks after the first pinch. Each shoot should then produce 2 further shoots that flower in due time but later than the main flush. This results in a bigger production and a better continuous of the production. For the double pinch, all the lateral shoots are pinches off.

Factors

Land Preparation

Use raised bed for cultivation. Prepared raised bed of 15-20cm height, 1-1.2m width and of convenient length. Keep distance of 45-60cm between beds.

Sowing

- ☆ Time of sowing: Under greenhouse controlled environment, it can be grown throughout the year. For northern plains, planting is generally done in month of September-November and flowers are harvested in February to April.

- ☆ Spacing: For planting purpose suckers are used. Planting can be done on top of raised bed with spacing of 15cm x 15cm or 20cm x 20cm. Keep distance of 45-60cm between two beds.
- ☆ Method of sowing: Direct sowing of suckers on raised bed.

Seed

- ☆ Seed Rate: For one acre near about 75,000 suckers are required. Within 21 days, the cuttings normally develop good root system.
- ☆ Seed Treatment: Before planting, treat suckers with NAA@1000ppm (1gm/Ltr of water) it will help to improve root formation.

Fertilizer

Fertilizer Requirement (gm/plant)

UREA	SSP	MURIATE OF POTASH
67	125	17

Nutrient Requirement (gm/plant)

NITROGEN	PHOSPHORUS	POTASH
30	20	10

For good growth of crop, apply proper amount of fertilizer. Apply well decomposed cow dung@5-10kg at time of land preparation. Apply dose of N:P:K@30gm:20gm:10gm in form of Urea@67gm, SSP@125gm and MOP@17gm in green house.

Weed Control

To control monocot weeds, use Glyphosate@300 g/acre and for dicot weeds, apply Oxyfluorfen@200 g/acre as pre-emergent treatment.

Irrigation

Frequent watering is needed for carnation crop. Apply water 2-3times in a week during summer whereas in winter apply water for 2-3times with 15days interval. Apply water immediately after planting.

Harvesting

When to harvest is depend upon bud size and petal growth. Carry out harvesting in morning. Cut should be given on stem with help of sharp knife with care to avoid injury to stem or mother plant. In production period, cut flowers every two days. After harvesting of flower, immediately keep them in water or in preservative solution (Citric acid + Ascorbic acid@5ml each/lit of water) for atleast 4hours. Avoid exposing harvested flower to direct sunlight.

Post-Harvest

After harvesting complete grading operation and then bunch them properly. Flower are stored in cold storage at temperature of 2-4oC.

4.5.4 Lilium

Introduction

Lilium belongs to Lilleaceae, and are believed to have emerged 55 million years ago and has been in culture for the last 3,000 years (plot of lilies made its entry and was very popular in Egypt's 18th dynasty). Many bulbous plants called lilies. Many of them however belongs to other genera. The here mentioned all belongs to the genus Lilium. All modern lily hybrids derives from approx. 100 lily species, of which approx. a dozen tribes from Europe, approx. two dozen from North America, approx. 40 from China and the remaining approx. 25 come from Japan, Nepal, Myanmar and Korea. Lilies can be found only in the northern hemisphere.

Lilium

Botanical Description

Lilies appear in all shapes and sizes. All parts are divisible by three and some of these parts are twofold. The flower has three inner and three outer petals. There are six anthers, stigma is three-fold, and so is the ovary, each with two chambers.

Flower

- ☆ The flower associated with the stalk through a flower stalk can be funnel-shaped, spherical or backward bending. It can be upturned, outwards turned or downwards turned.
- ☆ The flowers may sit as an umbel where all flower stalks arises from a common centre.
- ☆ If the flower stalks do not arise from a common centre it is called a united umbel, or when all the flower stalks arise arbitrarily like the leaves, it is a raceme.

Leaf

- The leaves arise mostly horizontally from the stalk. In some case, when the leaves arise from the same centre, they are forming whorls.
- Many garden lilies have very similar leaves, but there are also variations, so if you have many lily species in the garden, there will be several fascinating leaf types for detailed study.
- Lily species as L. cernuum, L. pumilum and L. davidii and hybrid 'Elf' has many filamentous leaves emanating from the stalk.
- At the other end of the spectrum there are the lanceolate leaves with the widest part approx. halfway down the leaf from the tip, and the leaves narrows again. This is seen in e.g. L. auratum var. platyphyllum.
- Then there are a large number of leaf types, which lies between filamentous and large lanceolate.
- The leaves which are fastened in a wreath around the stem, usually are broad in shape.

Root

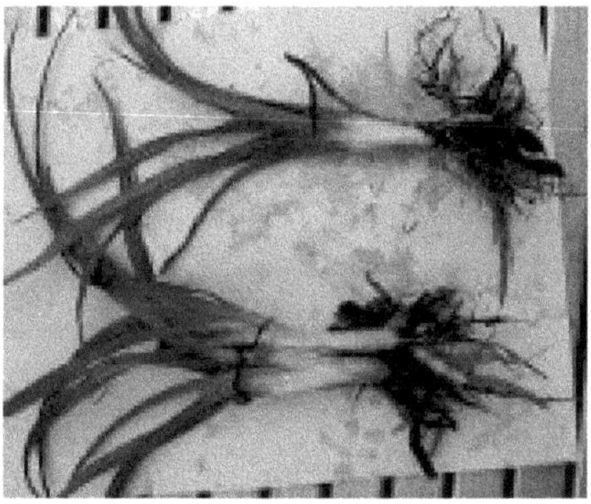

Lilium roots

- Throughout the growing season Lilium will develop a root system in the area below the surface and around the bulb. These roots serve to attach the stem into the soil so it does not fall over in wind and rain and when the lily blooms - often with heavy flowers. These roots also serve to record food and water for the plant.
- Lily has a bulb, which is composed of scales. These scales are transformed leaves that can sit more or less loose.
- The bulbs are usually white or yellow, but can be a little coloured when exposed to light. The size of the lily buds vary greatly. They can be as small as a grain or as large as a grapefruit. The greater flower the larger bulbs.

- ☆ The bulb has a flat area at the bottom called the basal plate. Lily will usually, in the growing season, form a system of roots that radiate out from this point. These roots are constricting and will pull the bulb deeper into the earth until it is satisfied with the depth. In some lilies works this mobility very well, while in others it goes less well.
- ☆ These basal roots collects important nutrient and moisture early in the season before the roots around the stem erupt.
- ☆ Throughout the growing season the lily will gather nourishment and energy for the next growing season. It will shape the new flower and embed it in dandruff. If the growing conditions are optimal, the lily will form more than one new flower and more than one new bulb for the next season.
- ☆ Some lilies, for example. L. canadense form offshoots arising near base. These stolons vary in length depending on the clone, its power and ground conditions. The stolon has small scales, which becomes a bulb, and it can have a good size at the end of growth period, when the stolon begins withering away.
- ☆ In a few North American types, e.g. L. pardalinum and its hybrids are formed rhizom-liking stems. A broad basal plate is formed on which the upturned scales are located.
- ☆ Scales of this type are very fragile, and are often narrowed in the middle so they break in half. Use any a sharp knife by dividing.
- ☆ A few Asian species forms stolons on which breaks one, two or three bulbs. These new plants can easily emerge 30-60 cm from the mother plant. This group includes L. duchartrei.
- ☆ Some lilies form bulbils (small bulbs) from the stalk.

Liliumin the Wild

- ☆ Lilies grow under very different conditions in nature. They are found on dry grassland, close to rivers, literally in swamp, in alpine canyons, on limestone mountains and rugged mountain terrain. There are lilies only growing in acid soil and others in lime soil, while others grow best in neutral soil.
- ☆ Some lilies have very limited distribution such. L. regale which is only found on a steep Chinese slop, while, for example. L. martagon is found from Siberia across to Poland and down to the Balkans.
- ☆ Differences in generic sizes and colors are often caused by growing conditions. One example is L. formosanum which grows in Taiwan in 0-3500 m. The lilies that grow in the lowlands may be 180 cm tall with numerous flowers, while those that grow high only are 25-30 cm and usually only get one single flower, between these extremes are many stages.

Lilium in Culture

- ☆ Most lilies require a well drained soil. Heavy soil and soil with standing water is fatal to lilies. Therefore, low-lying areas and areas where water collects and heavy clay soils, must be avoided.
- ☆ If the soil is too heavy, add sand, gravel or similar material to drain the soil.
- ☆ In nature, many wild lilies are covered by snow in winter and thus kept free of precipitation. In the spring rain and meltwater will bring the lily start growing. In culture, we must try to minimize water flow in winter and then maybe add extra water in spring and summer. By watering leaves, stems and flowers must be kept dry. By planting lilies between shrubs and similar (so that the Earth is covered) can prevent dehydration and thus reduce watering needs. This however should be weighed against the fact that lilies likes air around the plant.
- ☆ Lilies prefer a cool soil for the bulb. Cover the soil with pine needles, bark or similar. PH mostly should be around 6.5.
- ☆ Lilies are remarkably impervious to both heat and cold, but new shoots can be damaged by late frosts.
- ☆ Fertilizer shall be applied twice a year. First time on 15 April by mixing fertilizer 14-14-14 and the second time on 15 May with 11-22-25.
- ☆ It is best to plant lilies in autumn, so the basic roots may develop within the new growing season.
- ☆ Lilies should be planted 10-15 cm deep, because the stem-roots will develop on the lower approx. 10 cm of the stalk.
- ☆ Some believe that the size of the bulb is irrelevant when talking about the depth of planting, while others believe that small bulbs not shall be covered as much as big bulbs.
- ☆ There are however a few exceptions, such as. Lilium candidum, where the bulb should be placed in the surface or covered only slightly.
- ☆ When the bulbs after a few years will be located too close, the lilies will be smaller and weaker, therefore dig up the bulbs and divide them in appropriate intervals.

Propagation

Scale

- ☆ A simple way to propagate lilies is to take the scales from the bulb.
- ☆ Scales is usually taken from the bulb in the fall, but it can also happen at other times. By Asiatic hybrids scales can be taken in September, when

division normally occurs. If taken scales in August they will develop small bulbs before November when the rest period begins.

- ☆ When the bulb is exhumed rinse it in a fungicide. Brake the scales gently from the parent bulb as close to the basal plate as possible, preferably with a bit of the plate if possible.
- ☆ Do not use the outer dry scales, but only the thick-claims scales.
- ☆ The scales can be planted directly in a pot with moist soil or placed in a plastic bag with moist sphagnum peat moss. The bag is placed in a warm place for approx. 2 months. Check the bag regularly to ventilate and keep an eye on the moss not to dry out.
- ☆ When the first leaf emergence it is time to pot the scales. However, it has no hurry, since scales can survive several weeks.

Bulbils

- ☆ The lilies which forming bulbils can be propagated very simple by peel of the bulbils in late summer and place them in a plastic bag with damp moss. When the bulbils have formed roots it is time for potting. The bulbils can also be potted directly after harvesting in moist soil. Place the pot shade and take care it don't dry out.

Small bulbs

- ☆ Many lilies form small bulbs just below the surface. These small bulbs are in many respects similar to the previously mentioned bulbils. They're both formed on the stalk and they form their own basic roots. Propagation is done by picking the small bulbs from the mother plant and pot them in a moist soil mix in the shade.

Seeds

The above mentioned propagating forms gives plants which are identical to the parent plant.

A fourth way of propagation is by seed. But in this way the new plant will not necessarily be identical to the parent plant. In return, you may, if you like pollination, produce all sorts of lucky and unlucky hybrids. When the seed pot opens the seeds are matured. Harvest the seed and plant in a pot, either immediately or dry the seeds and store until December-January months and then do the potting. Set the pot outdoors in the shade. Some seeds sprouts immediately - so-called epigeale - (trumpet, Asian and Chinese), while others - the hypogeale - requires alternating hot and cold periods in order to germinate. They form initially a little bulb with a primitive root, then they require a cold period, sometimes several, before the increasing continues, and they shoot a leaf up from the ground. To the latter group belongs martagon, the oriental and most if not all American. It may take up to seven years before these lilies are in bloom.

Diseases and pests

The greatest scourge of lilies are undoubtedly the lily beetle. A small red guy who lets itself fall to the ground when disturbed. Here it will land on its back and is becoming difficult to find as it is black underneath. An effective remedy for lily beetles is to pick them by hand and kill them. There also is a systemic agent in the trade, which means that beetles do not like the taste of the plant, and they disappear. If we don't control the beetles on one way or another, we must pay the price later into the summer, where the beetles have left their larvae on the lily leaves. Larvae are wrapped in the excrement of the beetle, and it's a pretty disgusting task to assemble them. The larvae eat the lily leaves.

Lilium disease

One must control the lily beetle from early spring to autumn. Another problem is the larvae or worms that eat the bulbs. Here is so far known only one solution: dig up the bulb, remove the pest, plant the lily in a pot and care it until recovering from the attack. Fungus diseases are one of the most prevalent and there are many forms. E.g. gray mold (Botrytis) that starts with brown spots on leaves, which can fade completely into the summer. Botrytis only attacks the leaves and not the bulbs, but early wilting of the leaves and stalk, can make the bulb weak.

The supply of fungicides on the market is severely limited. Experiments with the available, and preventive - and by milder attacks try Atamon (a remedy used in the housekeeping for preserving). Ensure that the lily has ideal growing conditions, such as good drainage, damp soil, right fertilizing, etc, which might keep fungus away. If the bulb is attacked, you can dig it up, put it in a fungicide a few hours, let it dry slightly and then plant it in a good lily soil, preferably sterile.

Some species and varieties can be attacked of virus, but it is not always damaging the plant, and sometimes it will not be shown. If the bulbs are infected, it often means the plant dies. Destroy the infected parts to avoid infection. Some lilies are say to be particularly susceptible to be attacked of viruses such as L. longiflorum and L. formosanum.

4.6 PRODUCTION TECHNOLOGY OF ORCHIDS UNDER PROTECTED CONDITIONS

4.6.1 Introduction

Orchids are highly valued for their cut flowers in commercial floriculture owing to the wide range of colours, shapes, sizes and fragrance they display, with long vase life being an added advantage Orchids are monocots belonging to the family orchidaceae. Orchids are found all over the world, except in the arctic region, where the harshest clirnate of permanent frost or unrelieved aridity prevails. Although some Orchids are found in steamy jungles, most of them grow in far more pleasant climates. In fact, Orch~d family is one of the most widely adapted of all plant families. Orchids of one specles or another grow all over the world, except in the most severe arctic and sultry desert regions.

These plants differ from other flowering plants due to their pollen (pollina) that is formed into a mass, their stamens and pistils, which are joined together in a structure called a "column", and their seeds, which are very small, do not contain endosperm and have no organized embryo. Orchid flowers consist of the pedicel (the stem of the flower which includes the ovary), the sepals, the petals and the column. Sepals and petals are in threes. The three sepals consist of the dorsal sepal and two lateral sepals. The three petals consist of two petals and a modified one called the labellum, or more commonly, the lip. Orchids differ from other plants by the way they grow in nature. Most orchids are epiphytes and they grow attached on other plants (usually trees). Some others are semi-terrestrials that is they grow on the ground, on decomposing plant material .A small number of orchids are true terrestrials, meaning they grow in soil like most plants. Some orchids are lithophytes, which grow on rocks.

Orchids have two distinct patterns of growth - monopodial and sympodial. Monopodial orchids, such as Phalaenopsis and Vandas, have a single growth, typically an elongated stem that may be fairly short (Phalaenopsis) or reach several feet long like some Vandas, Renantheras or Vanilla. Sympodial orchids, such as Cattleyas, Oncidiums, Cymbidiums, Paphiopedilums, have several, sometimes dozens of successive growths. In sympodial orchids the life of each growth is determinate.

Varieties of Orchids

Perhaps the most intriguing feature of Orchid flowers is their striking resemblances to various forms of animal life. There is the Bee Orchid, the Moth Orchid, the Butterfly Orchid, the Scorpion Orchid, the Spider Orchid, the Lizard Orchid, the Frog Orchid, the Dove Orchid, the Tiger Orchid and even the Man Orchid. Not being satisfied with simply modeling their floral parts after several of these animal shapes, the Orchid seems to have tried their hand at caricaturing funny action posters of some animals. The British Monkey Orchid, Orchid Simia, has a lip, which captures the several postures of frolic of a playful monkey. The flower of the Donkey Orchid of Australia instantly brings to our mind the picture of stubborn donkey's head. Soldier Orchid, Orchids Militaries, gives us the unmistakable impression of a stuffy soldier earnestly on the march.

4.6.2 Factors affecting orchid growth

Light and temperature

Light and temperature are positively correlated. Orchids grow best in diffused light but there are certain orchids, which grow well under bright light. Orchids will tolerate levels of light on the higher end of their range provided the plant receives more water, more fertilizer and, if possible, better air movement to keep their leaves a little cooler. If the orchids get too much light and not enough water and fertilizer, they will be stressed. Visible signs of stress are shriveling pseudobulbs and / or leaves, drying buds and prematurely wilting flowers. If the leaves stay green, are crisp and firm, then the light intensity is right. If the foliage is dark green, then the light is too low.

If the foliage shows purplish marks or coloration, then the light is probably too high. Sometimes if the light is too high the tips of the leaves will dry up. Very few orchids will tolerate direct sunlight, except maybe for an hour or too after sunrise and an hour or two before sunset. In nature orchids inhabit various habitats, from low land in forests to mountain slopes, to swampy areas. Therefore some orchids need the warm conditions of a tropical rain forest, others need the cool conditions of a slope on the foothills of mountains and yet others to have needs somewhere in between.

Orchids are classified in 3 temperature groups:

- ☆ orchids that require cool temperatures
- ☆ orchids that require intermediate temperatures
- ☆ orchids that need warm temperatures

Water quality and watering

Good quality water is a very important requirement to grow orchids successfully. Acceptable water sources include deionized distilled water, tap water free of salts and rainwater. Watering needs depend on the potting media and the growing conditions. Ideally plants should remain highly moist and should never be dry. The quality of water is extremely important for good culture. In nature plants are drenched by rainwater. Rainwater is slightly acidic with a pH factor of 6.4 to 6.8. Hard water will create hard deposits on the leaves of plants.

This may clog the pores on the leaves and reduce transpiration. If the water is hard it may be beneficial to periodically clean the leaves with distilled water. Water may get in between leaves or new growth. If this water stays there overnight, when temperatures become cooler, it may promote the growth of bacteria and fungi that may kill the new growth. To reduce risks of this happening, water only on sunny days. Watering early in the day will allow any water that got in between leaves or new growth to evaporate before night.

Humidity levels and air movement

Most orchids require 60 % to 80 % humidity. These humidity levels are necessary for the plants to perform their best and bloom. A humidifier will help considerably in maintaining the humidity and health of the plants. Another way of increasing humidity is by setting the plants on trays filled with pebbles or gravel and with water, taking care that the plants are not in contact with the water. Ideally a dozen or more plants should be grouped together as they will create a microclimate with higher levels of humidity. In the orchid houses there should be continuous air circulation.

4.6.3 Propagation Method of orchids

There are three main methods of propagating orchid plants:

- ☆ Macrovegetation which involves dividing large plants, removing offsets and plantlets;
- ☆ Microvegetation by meristem or mericlonal culture by growing and dividing small growing tissues in culture media
- ☆ Seed

Vegetative propagation by division

Many species of epiphytic orchids, especially Dendrobiumand Epidendrum, produce small plantlets on their stems and these can be removed and potted up. Other pseudo bulbous species, such as some Pleiones, often produce small offset bulbs, which after perhaps several years of careful culture will reach flowering size. Some orchids grow by developing new growth from the base of the plant. After several years they may have 5 to 10 or more growths, which may be subdivided.

The usual way of propagating is by dividing large plants. Long stemmed monopodials such as species of Vanda, Renanthera and Aerides are among the most difficult to divide. Only older healthy plants that have developed aerial roots 20cm or more from their bases should be propagated by cutting them into pieces below each root.

Propagation by seed

Growing orchids from seed are a delicate and complex activity.

In vitro seed culture

Orchids seeds are extremely small and usually undifferentiated. They are produced in large numbers. Each seed contains an undifferentiated embryo composed of 80-100 cells without any functional endosperm. At IIHR, a technique was developed for viability testing of orchid seeds with the help of tetrazolium chloride (2,3,5 Triphenyl 1-2-4 tetrazolium chloride) and malachite green.

The chemical, tetrazolium chloride is imbibed by the seeds as colourless solution and is reduced by the enzymes present in the viable embryo to a red coloured substance formazan. In the absence of any enzyme the dead embryo will remain uncoloured composition of nutrient media. Different media are used in orchid seed culture. Some media like Knudson C, Vacin and Went are broad spectrum, while media like Burgeff are exclusively for Paphiopedilum seeds.

Green pod culture or embryo culture

In green pod culture process the seed capsule is removed from the plant after fertilization. The surface of the capsule is cleaned with sterilizing agent and flamed after dipping in alcohol; the seed capsule is opened with a sharp blade and seeds sown directly into medium under aseptic conditions. The embryo on the nutrient media start turning green after 10-20 days depending upon genera. It is recommended to prick out the seedlings 2-3 times in vitro conditions before they are big enough to be transplanted. The medium should be changed with each pricking, reducing the level of sucrose when seedlings are big enough (3-4 months), they can be taken out.

Tissue culture

Vegetative propagation method, meristem culture, has gained importance. However, mersitem propagation has only been achieved, to a commercially acceptable standard, in Cattleya, Cymbidium, Calanthe, Dendrobium, Miltonia, Odontoglossum, Oncidium, Zygopetalum and hybrids. Plants produced by this method, when young are known as mericlones. Here the meristems are transferred to a culture medium where they can be subdivided and an unlimited number of new plants raised.

Apical meristem of a young shoot is the most commonly used explant both in monopodial and sympodial orchids; terminal and axillary buds from shoot tip containing meristem have also been used in many cases. The other parts of the plant used are inflorescence segments, root tip, lateral buds and leaf. Orchids can easily proliferate on very simple medium like Knudson C or the more stable Vacin and Went medium. Several adjuncts like coconut water (150 mg/l), banana pulp (100 g/l) and peptone (1-2 g/l) are commonly used.

4.6.4 Containers for growing orchids

Orchids can be cultivated in different ways. Among the various methods of cultivation, cultivation in coconut husks, in clay pots and on grounds, at open places under shady trees or in greenhouses. Cultivation in coconut husks is relatively cheaper when compared to other methods of cultivation since the medium holding the plants are coconut husks which are locally available at a very low price. Large-scale commercial cultivators very often prefer to cultivate Orchid plants in 6" and 8" pots, having sufficient air holes to enable aeration to the plants.

Even though 8" pots are costlier than 6 pots, it has got several advantages over 6 pots. Big sized pots enable luxuriant growth of the plants resulting in increased number of flowers having bigger sizes when compared to other methods. Unlike smaller pots, this method of cultivation does not require frequent re-potting which is a costly affair to a commercial cultivator. Besides these, relatively larger numbers of kiekies are obtained from Orchid plants grown in larger pots. Moreover, cultivation in clay pots can be carried out in any place irrespective of the topography of the region. Hence cultivation of Orchid plants in clay pots is more popular in the waterlogged areas of the State.

The most common containers / supports for growing orchids are:

Plastic pots

Most commercial growers use plastic pots because they are inexpensive, they are lighter and they are easier to store; hold water for a longer period than other pots / containers, mineral salts (from water and fertilizer) will not adhere to them, and roots will not get attached to them. Plastic pots are excellent containers for growing orchids. Pots with a number of drainage holes have to be used (4 to 8 holes on 3" to 4" pots, 8 to 12 holes on 5" to 6" pots).

Clay pots

The advantage of clay pots is the stability due to their weight and their porosity, which allows the potting material to dry faster. Of course, drying faster can also be a disadvantage. A possible disadvantage is that roots tend to attach themselves to the pot.

Clay orchid pots

Have either holes or slits on their sides to allow more air circulation than regular clay pots. As a result, they dry even faster than regular clay pots.

Vanda baskets

Vanda baskets are used mostly for Vandas and vandaceous orchids, but can be used for most orchid genera. Most Vanda baskets are made of teak wood.

Cork slabs

Cork slabs are used for mounting orchids. Pieces of cork range can be as small as 2" by 3" or as large as 12" by 24". Some orchids will only thrive when mounted on a piece of cork or on a tree fern slab or a piece of driftwood, but many that grow fine in pots will also thrive on a piece of cork and it makes for a much more natural look.

Driftwood

Driftwood is an alternative for cork that also can make for some very intriguing and interesting compositions.

4.6.5 Potting orchids

Potting monopodial orchids

The plastic pot size should be just large enough to accommodate the roots. If a clay pot is used, a pot that's just one size larger than the plastic pot, allowing for about 1" of space around the roots. The plant is centered and held so that the junction of roots and lower leaves is in line with the top of the pot, and potting mixture is added. While holding the plant from the base, the potting mixture is firmed to the inside rim of the pot.

When done the base of the plant should be just a little higher than the media so that leaves do not touch the media and the top of the roots are just a little bit exposed. If necessary the plant is staked so that it does not topple.

Potting sympodial orchids

The procedure for potting these is instead of centering the plant the older portion of the plant should be as close to one edge of the pot, leaving room on the opposite side of the pot for the new growth(s). When done the base of the plant or the rhizome should be just a little higher than the potting media so that leaves do not touch the media and the top of the roots are just a little bit exposed.

Repotting and care after repotting

Repotting is done after 2-3 years if the plants have over grown the pots. The old roots and leaves can be pruned. Plants can be divided with a minimum of 3-4 shoots and a sprouted bud. Repotting is done using a fresh media. Repotting is a shock to plants. So for a few weeks after repotting a plant must be carefully nurtured. For 3 to 4 weeks the newly repotted plants must be placed at a location where they will be getting less light than what they usually get. The lower light levels will reduce the stress caused by the repotting shock and will help the plants recover better and faster. Water lightly (just enough to get the potting material moist, not enough for the water to run through the drainage holes) until a week after repotting.

4.6.6 Fertilizing orchids

Three basic principles apply to orchid feeding:

- ☆ Only well rooted plants should be fed
- ☆ The feed is preferable in small but frequent doses – It should only be applied during the period of maximum active growth. A fertilizer complex containing NPK in equal proportion (17:17:17) ranging between 0.2 to 1.0% is applied depending on the stage of the crop. During vegetative phase, 30:10:10 combination may be used which may be changed to 10:20:20 during blooming stage. One of the liquid sprays which are generally used is the Ohio WP solution and the composition is:
- ☆ Potassium nitrate: 2.63 g
- ☆ Ammonium sulphate: 0.44 g
- ☆ Magnesium sulphate: 2.04 g
- ☆ Monocalcium phosphate: 1.00 g
- ☆ Calcium sulphate: 4.86 g
- ☆ Iron sulphate: 0.5 g
- ☆ Manganese sulphate: 0.25 ml

This spray mixture is made up for 1 gallon (4.5 litres) of water. Organic manure like cowdung, neem oil cake, poultry manure and so on soaked in water for 4-5 days for fermentation and subsequently diluted 10-15 times may also be used. Coconut water spray also is advocated. Generally the frequency of application is twice a week under tropical conditions. Nutrient availability to plants is affected by pH levels. Most nutrients are available at a pH between 5.5 and 6.5 for orchids.

Yield

The yield is variety specific and on an average, about 5-7 spikes per plant in the second year and 6-8 spikes per plant from the third year onwards can be obtained.

4.6.7 Harvesting

Harvesting should be done in general when most of the flowers on the spike have opened fully with two or three buds. Flowers cut prior to maturity wilt. In Dendrobium a spike can be cut at last 1or 2 buds are yet to open or 20 per cent of flowers are in bud condition.The stalk is cut at the base close to the stem. This operation is generally done in the evenings and the stalks have to be pre-cooled in water before packaging and storage.

Immediately after harvest, the lower 0.75cm of the peduncle is cut off, and the flower is inserted into a fresh tube of water containing preservative. Treatment with 8-HQC+5% sucrose improves the flower quality and vase-life of flowers.

Stage of harvest

Stage of harvesting is important, it have influence on the keeping quality of spikes. Dendrobium spikes are harvested while two or three buds are still unopened, it adds to the charm of the spikes. This applies to all other orchids, which produce long spikes like Phalaenopsis, Aranda, Arachnis, etc. The number of unopened buds to be retained may be lesser in smaller spikes.

Method of harvest

The spikes that are ready for harvest are cut at the base of the stalk, close to the stem. The cut surface should be smooth and should never be crushed. A sharp knife or secateur can be used for this. When several spikes are cut on the same day using a single knife, there are chances that the diseses are transmitted through the sap. The cut surface may act as substrate for disese causing organisms. Dipping the knife in an antibiotic-fungiside mixture and pasting the mixture on the cut surfaces can help in the prevention of this. Aternatively, flaming the cut surface can also be done. This will help in the sealing of the wound too.

Post harvest operations

The harvested flowers are immediately dipped in a bucket of water till they are packed. When the spikes are cut from the plant there is chance of entry of air through the cut surface of the spike.

If another cut, removing a small portion from the base of the spike, is made under water, entry of air can be prevented. In order to improve the quality of flower spike and increase in the vase life pulsing treatments are given to the cut spikes. In Dendrobium hybrids pulsing with sucrose 6% or BA 75 ppm improved vase life of flower spikes significantly. Spikes of Dendrobium pulsed with HQ 500 ppm + sucrose 5% for 6 hours recorded the longest period of vase life, decreased the respiration rate, improved bud opening and increased water uptake.Cotton dipped in water and placed around the base of the flower stalk A piece of polythene sheet is then kept in place around the base of the stalk using a rubber band. Small narrow bottles filled with water can also be used. The stalk end is inserted into the bottle and sealed with a cellophane tape around the stalk and the mouth of the bottle.

4.7. PRODUCTION TECHNOLOGY OF GLADIOLUS, TUBEROSE, CHRYSANTHEMUM UNDER OPEN CONDITIONS

4.7.1 Gladiolus

Introduction

The gladiolus was once known as sword lily. Improved types were first developed from crosses of several species native to the Mediterranean area of Europe. Later discovery of African species led to crosses, which produced the forerunners of the attractive large-flowered types we know today.

Glads

Although "glads," as they are commonly called, are used to a limited extent for landscape effect, their chief value is for cut flowers. Their wide range of colors, sizes and flower types makes them particularly useful for flower arrangements They are spectacular flowers for exhibition in flower shows and are a specialty of many amateur growers.

Types

Although the specialist may be interested in growing some of the unusual species and classes, the home gardener is concerned basically with two types of glads: the large-flowered and miniature.

Large-flowered

These varieties with their large, showy flowers are the most popular of all gladioli. They are excellent for exhibition and are much used by florists as cut flowers. Petals may be frilled, ruffled, semiruffled or plain. Almost all colors, tints and shades are available in them.

Miniature

The miniature glads should not be confused with varieties of species such as Primulinus. They are not plain-petaled and small but are essentially replicates of the ruffled and frilled large-flowered types. The small flowers on short spikes often make them more suitable for small arrangements.

Classes

Gladioli are classified into five categories to indicate size. In the code, the first number indicates size (Table). The diameter of the flower is measured as it grows normally without being flattened out in any way. This measurement is made on the bottom florets of the spike, which are generally slightly larger than those farther up.

The middle and third numbers of the code signify color (Table). An even number for the third digit means no conspicuous markings. An odd-numbered third digit means that the flower contains conspicuous markings.

The number 454, for example, indicates a large-flowered medium-red glad without prominent markings, while 262 indicates a small-flowered glad with a light rose color.

Table: Gladiolus size classes

Code	Flower diameter
500 (Giant)	over 5-1/2 inches
400 (Large)	4-1/2 to 5-1/2 inches
300 (Medium)	3-1/2 to 4-1/2 inches
200 (Small)	2-1/2 to 3-1/2 inches
100 (Miniature)	under 2-1/2 inches

Table: Gladiolus color indicators

Color	Pale	Light	Medium	Deep	Other
White	00				
Green		02	04		
Yellow	10*(1)	12	14	16	
Orange	20*(2)	22	24	26	
Salmon	30	32	34	36*(3)	
Pink	40	42	44	46	
Red		52*(4)	54	56	58 Black red
Rose	60	62	64	66	68 Black rose
Lavender	70	72	74	76	78 Purple
Violet (Blue)	80	82	84	86	
Smokies		92	94	96	
Tan	90				98 Brown

*Includes (1) Cream, (2) Buff, (3) Orange scarlet, (4) Red scarlet

Gladiolus bulb

The gladiolus "bulb" is actually a food-storage structure known as a corm. Gladioli growing natively in South Africa were exposed to a dry season. It was the function of the corm to maintain the plant while dormant until growth resumed after the spring rains began.

The corm is a swollen underground stem. Each year a new corm is formed atop the old one, which shrivels and dies. On the upper surface of the new corm, buds develop from which the new plant grows the following year. The bases of old leaves are papery and cover the corm. These are called husks. The husks overlap each other and meet to form a point at the top.

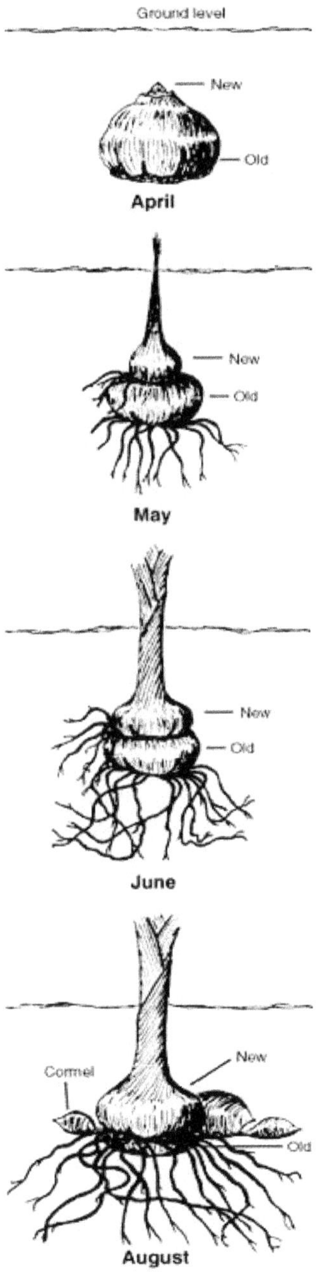

Development of gladiolus corm

While the new corm is forming atop the old one, small new corms called cormels or cormlets are produced from the base (Figure). Cormels are a chief means of propagating a certain variety of gladiolus.

Selecting corms

Healthy corms are essential for producing vigorous plants with flowers of maximum size for each variety. A healthy corm should have smooth husks that are not discolored or damaged. The corm itself should be firm and without any dark areas or spots on the surface after the husk is removed. The base of the corm, where the old corm was removed, should be firm.

The size of the plant and spike produced is directly related to the size of the corm that is planted. Corms are graded into sizes as follows:

- ☆ Jumbo: 2 inches or more in diameter
- ☆ Number 1: 1-1/2 to 2 inches in diameter
- ☆ Number 2: 1-1/4 to 1-1/2 inches
- ☆ Number 3: 1 to 1-1/4 inch
- ☆ Number 4: 3/4 to 1 inch
- ☆ Number 5: 1/2 to 3/4 inch
- ☆ Number 6: 3/8 to 1/2 inch

Sizes Jumbo, 1, 2 and 3 can all be expected to flower (sometimes two or more spikes each) under good growing conditions. Occasionally under ideal growing conditions, a flower will be produced by sizes 4 and 5. Corms are most productive during their first six years. Often as they become older they may decline in vigor. For this reason it is desirable to save cormels to develop a constant supply of young, vigorous corms.

Cultural practices

Soil

Gladioli can be grown well in many different soil types. If there is any possibility of selecting soils, a loam or sandy loam should be preferred. Soils with a pH between 6.0 and 6.5 and with medium fertility will give the best growth. If the fertility condition of soil is unknown, a soil test will help determine levels and indicate any need for improvement. Soils that produce good potatoes, corn or other garden vegetables will produce good glads. The vegetable garden is a good place to grow glads for cut flowers.

Heavy soils with poor internal drainage can be lightened by adding organic materials such as compost, peatmoss or sawdust. If sawdust is used, nitrogen in the soil will be tied up. For this reason, it will be necessary to mix about 3/4 cup of ammonium nitrate with each bushel of sawdust. In areas where drainage is poor, glads should not be grown, or raised beds should be built to improve drainage.

To maintain the fertility in the bed, apply 2 pounds of a 5-10-5 fertilizer or one with a similar analysis to each 100 square feet of planting space. Scatter the fertilizer evenly over the soil surface before digging, and dig to a depth of 6 to 8 inches. Rotted cow manure is excellent for glads but must be mixed well into the soil so it does not touch bulbs or foliage, or it will encourage rots.

Planting

Select a spot in full sun with good drainage. Glads do not do well in competition with woody shrubs or trees and therefore should not be planted close to them.

Glads may be planted as early as a month before the average last frost date and at two-week intervals thereafter for a succession of bloom. The last planting should be no later than early July if the corms are to have enough time to develop and mature before frost. The time from planting until bloom varies from 70 to 90 or more days, depending on weather conditions and variety. Plant large bulbs 4 to 6 inches deep, medium-sized bulbs 3 to 4 inches deep and small bulbs 2 to 3 inches deep. Cover only 2 inches in the beginning and pull soil to the plants with each cultivation. Glads are often planted by digging out a trench and planting corms in either a single or double, staggered row. Corms may be spaced only 2 to 3 inches apart in the row, but if they are being grown for show, they should be spaced at least 6 inches apart. Rows should be spaced from 20 to 36 inches apart.

Cultivation

Cultivation of glads should be only enough to keep down competing grasses and weeds. It should be shallow so that roots are not damaged. Mulches help to keep down weeds and conserve soil moisture. Straw, wood shavings and sawdust are all suitable. A 2- to 4-inch depth is required for good weed control. A light application of nitrogen should be made after a heavy mulch is applied. Preemergence herbicides can be used successfully for the control of most annual weeds and grasses to reduce cultivation.

Watering

Glads need ample water throughout the growing season. During dry weather, water the plants weekly to supply the equivalent of 1 inch of rainfall per week. Watering should soak the ground thoroughly. Avoid daily light waterings.

Staking

To ensure tall, straight flower spikes, staking is necessary. In beds or borders, individual plants will need to be staked. This can be accomplished using 1 x 1-inch stakes made from any suitable lumber. The spikes should be tied to the stakes at about 10-inch intervals with soft twine or cloth strips. In the cutting garden (as the row in the vegetable garden) 2 x 2-inch stakes can be inserted at about 10-foot intervals along the row. Wire or heavy twine is then stretched between the stakes at a spacing of 10 and 20 inches above the soil line. Individual plants can then be tied to the supporting wire or twine. Hilling up soil on both sides of the row likewise gives good support.

Fertilizer side-dressing

A side-dressing of fertilizer after the plants are well established will increase vigor. Apply a 5-10-5 fertilizer to each row. This can be applied in a band on one side of the row, keeping 4 to 6 inches away from the stems. It will be most effective if applied in a small trench several inches deep but may also be placed on the soil surface and scratched in lightly. The first application should be timed when growth is about 6 to 10 inches tall. A second application is placed on the opposite side of the row as the flower spikes start showing through the leaf sheaths. Apply about 1 pound of the fertilizer to each 100 feet of row at each application. Too heavy applications of nitrogen early in the growing season result in excessive foliage growth and poor quality flowers.

Cutting flowers

Glads do not look well in the garden unless old flowers are kept picked. Since they are used primarily for cutting, spikes should be picked in their prime for maximum life indoors. Cut the spike when the first floret is showing color. To cut the spike, insert a sharp knife above the second to fourth leaf and make a slanting cut up the stem. Immerse in water immediately to prevent wilting of the petals. Be sure to leave at least two, and preferably four, leaves on the plant after cutting spikes to help corms mature properly. Gladiolus spikes are very geotropic and should be stored in a vertical position until they are used.

Digging corms

Corms are ready to dig in four to six weeks after blooms are finished or when the tops die off. They can be dug anytime before a hard freeze. Healthy plants should be left in the soil as long as possible so there is ample time for maximum development of the corms.

Curing and cleaning

After digging, wash off soil that adheres to the corm and roots. Cut the tops to within one-half inch of the corm. Corms can be left outdoors in the sun for a day or two if the temperatures are mild, and then spread out in a light, airy place to cure. They are cured to get the surplus moisture out of the husks and corms as quickly as possible to prevent storage rots. After two to three weeks of drying, remove the old corm from the base. Sort the corms and cormels according to size. The small cormels can be saved and planted the following year, but remember it will take two to three years to produce a blooming-size corm from them.

Storing

Corms should be stored during the winter at a temperature of 35 to 45 degrees Fahrenheit in a well-ventilated area. Airy containers such as loose-weave baskets, mesh bags or old nylon stockings make good containers that may be hung out of the way. Dust corms before storage with an all-purpose garden fungicide to ensure against insect and disease damage while in storage.

Storing corms

Store corms in a well-ventilated area in a loose-weave basket, mesh bag or old nylon stocking.

Insects

Gladioli are relatively free of insect pests. The most troublesome pest is the gladiolus thrips. This insect is very small and seldom seen as it feeds in hidden places. It does considerable damage, especially to the flowers. It works on the buds before they emerge from the sheaths and causes malformed and spotted flowers. Thrips may overwinter on the corms. Aphids, grasshoppers and cucumber beetles are other insect pests that may cause damage to flowers or foliage.

Diseases

There are many different disease organisms that attack the corms and stems of the gladiolus. Collectively, they might best be called corm and stem rots. Most are active during storage and develop with improper curing and storage. Before corms are planted in the spring, they should be carefully inspected. Pull back some of the husk to examine the inner surface and eliminate all corms that are infected. Although infected bulbs may grow if planted, growth is generally weak and stem rots often develop later to cause death. One of the best control measures against these rots is immediate removal and destruction of bulbs or plants that show disease.

4.7.2 Tuberose

Introduction

Tuberose is one of the most important tropical ornamental bulbous flowering plants cultivated for production of long lasting flower spikes. It is popularly known as Rajanigandha or Nishigandha. It belongs to the family Amaryllidaceous and is native of Mexico. Tuberose is an important commercial cut as well as loose flower crop due to pleasant fragrance, longer vase-life of spikes, higher returns and wide adaptability to varied climate and soil. They are valued much by the aesthetic world for their beauty and fragrance. The flowers are attractive and elegant in appearance with sweet fragrance. It has long been cherished for the aromatic oils extracted from its fragrant white flowers. Tuberose blooms throughout the year and its clustered spikes are rich in fragrance; florets are star shaped, waxy and loosely arranged on spike that can reach up to 30 to 45 cm in length. The flower is very popular for its strong fragrance and its essential oil is important component of high- grade perfumes.

Single varieties are more fragrant than 'Double' type and contain 0.08 to 0.14 percent concrete which is used in high grade perfumes. There is high demand for tuberose concrete and absolute in international markets which fetch a very good price. Flowers of the Single type are commonly used for extraction of essential oil, loose flowers, making garland etc., while that of Double varieties are used as cut flowers, garden display and interior decoration. Fragrance of flowers is very sweet, floral and honey-like and can help give emotional strength. The flower spike of tuberose remains fresh for long time and finds a distinct place in the flower markets. Due to its immense export potential, cultivation of tuberose is gaining momentum day by day in our country.

The scientific name of tuberose is Polianthes tuberosa. And it belongs to Amaryllidaceae family. Basically, it is half-hardy, perennial, bulbous plant. Bulbs come with scales and leaf bases and stem remains concealed within scales. Roots are mainly adventitious and shallow. Additionally, flowers have a funnel-shaped perianth and are fragrant, waxy white, about 25 mm long.

Basically, both the tropical and sub-tropical countries grow this flower. South Africa, North Carolina, USA, and India are the major growers of tuberose. In India, states like Karnataka, Andhra Pradesh, Tamil Nadu, Maharashtra, and West Bengal grow the flower commercially.

Tuberose can successfully be grown in pots, borders, beds and commercially cultivated for its various uses. The flowers of tuberose are also used for making artistic garlands, floral ornaments, bouquets, buttonholes, gajras and extraction of essential oil. It is also a popular cut flower, not only for use in arrangements, but also for the individual florets that can provide fragrance to bouquets and boutonnieres. The long flower spikes are excellent as cut flowers for table decoration. The flowers emit a delightful fragrance. Tuberose represents sensuality and is used in aromatherapy for its ability to open the heart and calm the nerves, restoring joy, peace and harmony. Tuberose flowers have long been used in perfumery as a source of essential oils and aroma compounds. Tuberose oil is used in high value perfumes and cosmetic products. Furthermore, fragrant flowers are added along with stimulants or sedatives to the favourite beverage prepared from chocolate and served either cold or hot as desired. Tuberose bulbs contain an alkaloid-lycorine, which causes vomiting. The bulbs are rubbed with turmeric and butter and applied as a paste over red pimples of infants. Dried tuberose bulbs in powdered form are used as a remedy for gonorrhoea.

Varieties:

There are four types of tuberoses named on the basis of the number of rows of petals they bear. They are,

- ☆ Single
- ☆ Semi-double
- ☆ Double and
- ☆ Variegated

Single variety

They bear pure white flowers with one row/ whorl of corolla segment. Flowers are highly scented and are extensively used for loose new.pngflower purpose, essential oil and concrete extraction. Single types are more fragrant than double. Concrete content has been observed to be 0.08 to 0.11 per cent.

Loose flowers are used for making floral ornaments. Its floral buds are greenish white. Also the per cent seed setting is high in single. Single Mexican, Kalyani Single, Shringar, Prajwal, Arka Nirantara, Rajat Rekha, Hyderabad Single, Calcutta Single, Phule Rajani, Kahikuchi Single, Pune Single are main varities.

Semi-double:

They consist of flowers bearing 2-3 rows of corolla segments on straight spikes. semi. pngsemidou.png

Double:

They bear flowers having more than three rows of corolla segments It is a single flowered variety. This on straight spikes. Flower colour is white and also tinged with pinkish red. The main varieties are Pearl for Double, Kalyani Double, Swarna Rekha, Hyderabad Double, Culcutta Double, Vaibhav and Suvasini.

Soil and Climatic requirement

Climate

Tuberose is best suited for cultivation in tropical to subtropical and temperate climates. Tuberose prefers to grow in an open sunny location, away from the shade of trees. It requires warm and humid climate although flowering is profuse under mild climate. In India,commercial cultivation of tuberose is confined to warm humid areas with average temperature ranging from 20° to 30°C. The crop is reported to flower profusely throughout the year, if the climate is mild and free from extremes of high and low temperature. If the temperature is above 40°C, the length of the spike and quality of the flowers are severely affected. Very low temperature and incidence of frost will damage the plants and flowers. Tuberoses grow well in a sunny situation and should not be grown on shady or semi-shady situation which drastically reduces the flower yield. Humidity and temperature are the two main factors responsible for increasing or decreasing the production of tuberose.Tuberose although not strictly photosensitive, long-day exposure promotes vegetative growth as well as early emergence of the first flower spike and also increases flower spike length.

Soil

Tuberose can be grown on wide variety of soils ranging from light,sandy loam to a clay loam. It can also be successfully grown as a commercial crop even in those soils which are affected by salinity and alkalinity conditions if better agronomical practices are adopted. The soil should be at least 45 cm deep, well drained, friable, rich in organic matter and nutrients with plenty of moisture in it. Tuberose should grown in well drained place. Crop is sensitive to water stagnation it cannot tolerate water logging even for a short period. So essential to ensure proper drainage pests. or else planting should be done on done on bund. Fertile, loamy and sandy soils having a pH in the range of 6.5 to 7.5 with good aeration and drainage are ideal for tuberose cultivation. A place protected from strong winds is the soil is preferable.

Selection of site

Tuberose is a sun-loving plant. For attaining good vegetative growth and maximum yield of flowers, a site should be chosen, where the plants will get plenty of sun throughout their growing period. The soil should have sufficient moisture holding capacity. A place protected from strong wind hot and cold waves is preferable, so that there is no damage of flower stalk. In shady situation, the plants grow tall and lanky and the flowering s adversely affected. They are highly susceptible to water stagnation and poor drainage conditions and hence require well drained and aerated soil.

Field or land Preparation

The land should be ploughed deep to a good tilth, 2-3 times, to a depth of 30 - 45 cm. The first ploughing should be done in January. The field is left exposed to sun for at least 15 days that destroys the weeds and insect-pests. The second ploughing can be done about a month before planting. At the time of the second ploughing, well rotten farmyard manure at the rate of 50 tonnes per hectare should be incorporated into the soil. Then, the soil is brought to a fine tilth by breaking the clods and removing the weeds. After field preparation, plots of Selection of site convenient sizes should be prepared with irrigation channels, ridges and furrows at the recommended spacing.

Season of planting

Well-developed spindle-shaped bulbs, with diameter 1.50 cm and above forming at the outer periphery of the clump, are considered ideal for planting. Tuberoses are generally planted in March - April in the plains and April-May in the hills. The bulbs can also be planted during July -August. To obtain flowers almost throughout the year, sequential planting can be practiced. Freshly harvested tuberose bulbs can be used for planting 4 - 5 weeks after harvesting. Planting fresh bulbs leads to profuse vegetative growth and poor flowering.

Propagation

Tuberoses are propagated by bulbs, bulblets and seeds. Multiplication by bulb-segments and in vitro micro propagation from scale stem-sections is also practiced.

Propagation by bulbs

Most common method practiced for the commercial multiplication of tuberoses is through propagation by bulbs. The bulbs remain dormant during the winter months in places where the temperature is low. The dormancy of the bulbs can be successfully broken by dipping the bulbs in 4% Thiourea solution for one hour if early planting is desired. Ethylene chlorohydrins can also be used for breaking the dormancy of bulbs. The bulbs are separated from the clumps by rubbing off the loose scales and the long roots should also be removed. Selection of suitable bulbs is very important for successful cultivation. In general, spindle -shaped bulbs free from diseases having diameter between 1.5 and 3.0 cm are suitable for planting. About 1.25 - 1.5 lakh bulbs (8 to 9 tons of bulbs) are required for planting one hectare.

Propagation by bulb segments

Propagation through mature bulbs is expensive, therefore, multiplication of growing stock can be done by division of bulbs. Large sized bulbs having 2.1 cm or more diameter are suitable for planting purpose. If the bulbs are very large, they are cut into 2 to 3 vertical sections, each containing a bud and part of the basal plate. Each of these sections is treated with copper fungicide and planted vertically in a rooting medium with its tip just showing above the surface. A moderately warm temperature should be maintained. New bulblets along with roots develop from the basal plate. At this stage, bulblets are transferred to the ground.

Spacing

Planting density markedly influences flower yield and quality. The planting distance varies with the soil and climatic conditions. Low planting density results in wastage of inputs and very high planting density leads more plant competition, thus reducing individual bulb enlargement For economic returns, bulbs are planted at an optimum spacing of 30 x 20 cm or 20 x 20 cm. About 40,000 - 50,000 bulbs are required for one acre planting of tuberose ie. about 1, 00,000 to 2, 00,000 bulbs are required for planting one hectare of land.

Depth of planting

The depth of planting varies from 3.0 - 7.0 cm depending upon the diameter of the bulb and the soil type.It should be 2.5 times more than the diameter of bulbs. While planting, the bulbs are planted at the recommended plant-spacing, 4-6 cm deep on the sides of the ridges. Planting is deeper in sandy soil as compared to clay soil.In sandy loam soil planting of bulbs is done at the depth of 6.0 cm. In general, planting is done in such a way that the growing portion of the bulb is kept at the ground level.

Seed / bulb treatment

Dipping the bulbs in 4% solution of thiourea can break the resting period. Pre-plant storage of bulbs at 10°C for a period of 30 days will improve the plant growth, increase spike and flower yield. Pre-planting treatment of bulbs with GA3, etherel or thiourea promotes early appearance of flower spike and produces longer spikes with maximum number of florets.The bulbs are first thoroughly cleaned and treated with Bavistin (0.2%) for 30 minutes. Dipping the bulbs for about 20-30 minutes in a solution of Emisan (0.2%), Thiram (0.3%), Captan (0.2%) or Benlate (0.2%) is also recommended. Dry in shade before planting or storing.Before planting treat bulbs in systemic fungicide and before storing in contact fungicide

Nutrient management

Tuberose responds well to the application of organic and manures. Apart from FYM (20 tonnes/ha), a fertilizer dose of 100 kg N, 50 kg P2O5 and 70 kg K2O per hectare is recommended for tuberose production.nutri.png For achieving increased essential oil content in flowers and for the maximum recovery of concrete, a fertilizer dose of 80 kg N, 60 kg P2O5 and 40 kg K2O per hectare has been recommended. Of the full recommended dose of fertilizers, half the N, the full dose of P and K has to be applied at the time of planting and the remaining half of N is given as a top-dressing after 45 days of planting.

Micronutrients

Apart from N, P and K, calcium,magnesium, sulphur, iron, zinc, manganese, aluminium, boron and copper have also been found to influence the growth and flowering in tuberoses.

4.7.3 Chrysanthemum

Introduction

Chrysanthemum are a member of the Compositae family and are available in a wide range of brilliant colors, shapes and sizes. First cultivated in China over 6 centuries ago, this type of daisy was initially grown as an herb associated with the power of life. The chrysanthemum flowers range from dazzling whites to deep bronzes, and the hardy plants are highlighted with full, dark green leaves.

Chrysanthemum flowers look like they have a multitude of petals, but each individual petal is actually a small floret. There are two different types of florets: ray and disc florets. Ray florets are what we traditionally see as the petals, while the disc florets create the center buttons. When the florets are all clustered together, they give us what we know and love as a mum bloom.

Types

With over 100 different chrysanthemum cultivars in the United States, the National Chrysanthemum Society has a classification system in place to categorize 13 different mums by flower shape.

Anemone

These daisy-like blooms feature long, tubular florets clustered around a tight button center. They form a 4-inch bloom in single or multiple colors.

Popular varieties include: Dorothy Mechum, Purple Light and Angel

Decorative

Florists use decorative class mums in floral arrangements. The 5-inch plus blooms have a flat appearance as the florets gradually get longer from the center out.

Popular varieties include: Fireflash, Coral Charm and Honeyglow

Irregular Incurve

Incurve blooms feature florets curving inwards. Irregular incurve mums feature large blooms between 6 to 8 inches. The florets curve in and cover the center of the flower. A few florets at the bottom of the bloom add fringe to the stem.

Popular varieties include: Luxor, Blushing Bride and River City

Intermediate Incurve

The florets of an intermediate incurve mum don't cover the center of the bloom. With shorter florets curving inwards, the less-compact bloom of an intermediate incurve only reaches a maximum 6 inches.

Popular varieties include: Apricot Alexis, Candid and Pat Lawson

Regular Incurve

Regular incurve chrysanthemum blossoms are tight, smooth globes of inwardly curving florets. Each bloom is between 4 to 6 inches in diameter.

Popular varieties include: Gillette, Moira and Heather James

Pompom

Resembling the regular incurve, Pompom mums are only 1 to 4 inches. The tight blooms are common in floral arrangements.

Popular varieties include: Rocky, Yoko Ono and Lavender Pixie

Quilled

Show-stopping quilled chrysanthemums feature long, tubular florets that open to a spoon shape or slight downward curve at the end. Their spiky appearance often mimics other types of mums.

Popular varieties include: Seatons Toffee, Mammoth Yellow Quill and Muted Sunshine

Single and Semi-Double

These daisy look-a-likes feature one or two rounds of ray florets around a compact center. Their total plant size is between 1 to 3 feet, making them ideal for small spaces and borders.

Popular varieties include: Rage, Icy Island and Crimson Glory

Spider

Spider mums are well known for their long, spiky florets of single or multiple colors. The tubular florets resemble spider legs and can go in all directions. The delicate and exotic appearance creates a focal bloom in your garden.

Popular varieties include: Evening Glow, Symphony and Western Voodoo

Spoon

Spoon mums have a button center surrounded by ray florets featuring a spoon shape at each tip. They are often mistaken for single chrysanthemums, but the difference lays in the slight curve.

Popular varieties include: Kimie, Fantasy and Redwing

Reflex

The bloom of a reflex mum is slightly flat with florets that curve downward. The crossing of the florets produces an interesting feather-like appearance.

Popular varieties include: White City, Champion and Apricot

Thistle

The thistle bloom, also called the bush bloom, often features multi-colored blooms. The long, thin florets twist to rise up or fall backwards towards the stem. Thistle blooms have a unique, exotic look to them.

Popular varieties include: Cindy, Cisco and Orange Spray

Unclassified

With so many chrysanthemum varieties, many chrysanthemum blooms feature characteristics that place them in more than one category. Unclassified mums exhibit a wide range of colors and sizes.

Popular varieties include: Lone Star, Lili Gallon and Pacificum

Factors required for plantaion

Taking the time to first understand how to plant chrysanthemums rewards you with full, beautiful plants loaded with blooms.

Time of Plantation

Planting chrysanthemum in the spring gives the perennial plant time to establish and adapt to its new garden home. You'll easily find mums in garden centers and nurseries in both fall and spring, but planning ahead is key to successful planting. It's tempting to buy those huge beautiful fall mums you see during the autumn season, but in terms of longevity, the smaller spring mums are actually a better investment.

The root system becomes stronger throughout the summer and fall, which increases a plant's ability to survive the winter. Planting in the spring will also result in a bigger bloom the following season. Although some fall mums can survive winter if planted immediately, the odds are much better with spring-planted mums.

Quality of Chrysanthemums

The chrysanthemums you purchase in garden centers are frequently referred to as "hardy mums" for a reason. The majority of mum varieties are winter hardy in Zones 5 through 9. Some varieties, such as Mammoth Daisy, are hardy down to Zone 3.

When shopping for mums, check the label to make sure you're purchasing hardy garden varieties appropriate for your planting zone. Local nurseries and garden centers will most often feature the varieties that are specific to local areas. Avoid purchasing from floral shops – their mums are different, less hardy floral varieties.

Soil

Chrysanthemums can survive in most soils, but they thrive in well-draining soil with consistent moisture. Growing mums in hard, dry soil prevents the roots from becoming well established, while wet, boggy soil drowns the roots. Finding the middle ground is key.

If you've planted other perennials, then you already know how to plant chrysanthemum. To create a good soil for your mums, work your soil to a depth of 8 to 12 inches. Mix in 2 to 4 inches of organic material, such as compost or peat moss. The perfect soil texture can be tested by taking a handful and squeezing. When you open your hand, the soil shouldn't clump or quickly fall apart. It should simply crumble.

The best suitable soil for chrysanthemum cultivation is well-drained sandy loam good textured soils. Having good amount of organic matter will result in excellent yield. Avoid the soils where too much of water stagnation is possible. The optimal soil pH range for its growth is 6.5 to 7.5.

Sunlight

Chrysanthemums are sun-loving plants. Although they technically require only 6 hours of sunlight each day, the more light they receive, the better their growth, bloom and hardiness. Slight shade in hot, summer afternoons is appropriate in warmer gardening zones to prevent scorching.

Mums bloom due to their photoperiodic nature. When the plant senses a change in the length of darkness in late summer, it begins to set buds. Planting near artificial lights, such as security lights or porch lights, can alter the bloom time of your mums.

Climate

Chrysanthemum plant is a short day plant, in other words, it requires long days for vegetative growth and short days for flowering. Light and temperature are the main factors that influence the growth of the plants and flowering. For its vegetative growth it requires long day with bright sunlight and high temperature ranging from 20 °C to 28 °C. For bud formation and flowering it requires short day and low temperature ranging from 10 °C to 28 °C. This crop requires the relative humidity of 75% to 90% which suitable for proper plant growth.

Land Preparation

The land should be prepared by giving 2 to 3 ploughings followed by harrowing to prepare the beds for planting. Supplementing the farm yard manure of 20 to 25 tonnes/ha in the last plough is recommended.

Spacing

It can be tempting to plant mums closely together. Smaller, spring mums don't seem to fill out a garden space as well. However, keep in mind that by fall, most properly planted mums will reach up to 3 feet in height and width. Like many perennials, mums will often become larger each year. Even if your flower bed looks a bit bare when you first plant your mums, in time it will fill in.

Spacing mums properly is essential for plant health. Plants that are too crowded compete for nutrients, have root system issues, attract pests and suffer from disease. Following the plant spacing directions for your chrysanthemum variety increases the health of your garden and protects your investment of time and money.

Propagation and Planting in Cultivation

Chrysanthemum plants are propagated vegetatively through suckers, terminal cuttings or by micro-propagation.

Suckers

In this method of propagation, after flowering, the stem should be cut back just above the ground level. This induces the formation of side suckers which should be separated from the mother plant and should be planted in sand bed. Apart from this, well rooted suckers can be directly transplanted in the field. Generally, 1,10,000 suckers should be required to cover the 1 hectare land. To protect against wilt disease, roots of the suckers should be dipped in 1 gram of emisan in one liter of water. These suckers should be planted during rainy season at 30 cm x 30 cm spacing on one side of the ridges. Pinching should be carried out once in a month after planting to induce more lateral branches on the plants.

Terminal cuttings

These terminal cuttings should be taken from a healthy stock plant. Length of cuttings should be about 5 cm – 7 cm. The cuttings should be dipped in 2500 ppm indole butyric acid (IBA) (rooting hormone). These cuttings should be put in sand beds in shade conditions.

Manures and Fertilizers

Chrysanthemum crop responds very well to manuring and requires about 10 to 12 tonnes of well rotten farmyard manure (FYM) per acre. This farm yard manure can be supplemented at the time of land or soil preparation. As a basal dose, apply 50 kg of 'N' (Nitrogen), 160 kg P2O5 and 80 kg K2O. For increasing the flower yield, spray GA3 at 50 ppm at 30, 45 and 60 days after planting. Micronutrients like foliar spray of ZnSO4 0.25% + MgSO4 0.5% can be applied. As part of biofertilizers, soil application of 2 kg each of Azospirillum and Phosphobacteria per ha at the time of planting should be applied. It should be mixed with 100 kg of farmyard manure (FYM) and applied.

Irrigation

The frequency of irrigation depends on the stage of growth, soil and weather conditions. Proper drainage system should be maintained for chrysanthemum grown both in beds and in pots. Irrigation should be carried twice a week in the first month and subsequently at weekly intervals.

Chapter 5

PRODUCTION TECHNOLOGY OF MEDICINAL PLANTS

5.1 INTRODUCTION

Global and national markets for medicinal herbs have been growing rapidly, and significant economic gains are being realized. According to the Secretariat of the Convention on Biological Diversity, global sales of herbal products totalled an estimated US$ 60 000 million in 2000. As a consequence, the safety and quality of herbal medicines have become increasingly important concerns for health authorities and the public alike.

Some reported adverse events following the use of certain herbal medicines have been associated with a variety of possible explanations, including the inadvertent use of the wrong plant species, adulteration with undeclared other medicines and/or potent substances, contamination with undeclared toxic and/or hazardous substances, overdosage, inappropriate use by health-care providers or consumers, and interaction with other medicines, resulting in an adverse drug interaction. Among those attributable to the poor quality of finished products, some clearly result from the use of raw medicinal plant materials that are not of a sufficiently high quality standard.

The safety and quality of raw medicinal plant materials and finished products depend on factors that may be classified as intrinsic (genetic) or extrinsic (environment, collection methods, cultivation, harvest, post-harvest processing, transport and storage practices). Inadvertent contamination by microbial or chemical agents during any of the production stages can also lead to deterioration in safety and quality. Medicinal plants collected from the wild population may be contaminated by other species or plant parts through misidentification, accidental contamination or intentional adulteration, all of which may have unsafe consequences.

The collection of medicinal plants from wild populations can give rise to additional concerns related to global, regional and/or local over-harvesting, and protection of endangered species. The impact of cultivation and collection on the environment and ecological processes, and the welfare of local communities should be considered. All intellectual property rights with regard to source materials must be respected. WHO has cooperated with other United Nations specialized agencies and international organizations in dealing with the above-mentioned issues. Such cooperation will be further strengthened through the development and the updating of relevant technical guidelines in these areas.

Safety and quality assurance measures are needed to overcome these problems and to ensure a steady, affordable and sustainable supply of medicinal plant materials of good quality. In recent years, good agricultural practices have been recognized as an important tool for ensuring the safety and quality of a variety of food commodities, and many Member States have established national good agricultural practice guidelines for a range of foods. However, quality control for the cultivation and collection of medicinal plants as the raw materials for herbal medicines may be more demanding than that for food production; possibly for this reason, only China, the European Union, and Japan have recently developed guidelines on good agricultural practices for medicinal plants. Since their guidelines were established to meet the requirements of specific regions or countries, they may not be universally applicable or acceptable.

5.2 CONTRIBUTION

Medicinal plants contribute significantly to the rural livelihoods. Apart from traditional healers practicing traditional medicine, more people than ever are involved in collecting, trad- ing and utilizing medicinal plants. The wide- spread use of medicinal plants in traditional medicines (TM) in developing countries including Nigeria is largely due to the diverse resource base, the flexibility, easy accessibility and affordability, particularly the rural poor. Medici- nal plants have therefore been considered a healthy source of life for the people who are able to harvest them from the wild to meet their primary health care needs. The result of the in- creased demand in both local and international markets as well as bio-prospecting activities searching for sources of new drugs is a source of great concern.

As estimated, about 95 % of Medicinal Plants in developing countries are harvested and collected in wild. As observed, foundation upon which the medicinal plant species and the traditional health care system survive is threatened by deforestation, population growth, urbanization. The difficulties experienced in the collection of plants from forest and the problems encountered there in together with heavy demand of medicinal plants necessitate domestication of medicinal plants. As reported, some protection given to medicinal plants could be achieved through regulation and the introduction of sustainable wild harvesting methods, but the domestic cultivation is an option that could both reduce pressure on wild populations and solve some of the problems inherent in the production of herbal medicines. It includes species misidentification, genetic and phenotypic variability, variability and instability of extracts, and toxic components and contaminants. Cultivation offers the opportunity to optimize yield, achieve a uniform, high- quality product and control in every stage of the production process. It might also allow us to modify concentrations of biologically important compounds through the manipulation of growing environments, through traditional selective breeding methods and through the application of modern biotechnology.

African Agricultural Production Systems

African farmers pursue a wide range of crop and livestock enterprises that vary both across and within the major agro-ecological zones. Food production and food security in Africa de- pend on many different systems, unlike other regions of the world where the contribution to food production and food security is based on a limited number of systems. Diversity is the norm in African farming systems. Even at the level of the individual farm unit, farmers typically cultivate 10 or more crops in diverse mixtures that vary across soil type, topographical position and distance from the household com- pound. The farming systems provide a snapshot of dynamic systems that are constantly evolving. Both endogenous factors (household goals, labour, technologies in use and the resource base) and exogenous factors (market develop- ment, shifts in demand, agricultural services and policies, the dissemination of new technologies and the availability of market and policy information) drive the evolution of individual farms and, collectively, the overall farming system.

Generally, farming system can be classified based on economic of production, water supply (rainfed or irrigated farming), system of production (collecting, ley, sole cropping, mix or multiple cropping). In the developing countries, agricultural production involved traditional subsistence with mono cropping, multiple cropping and livestock management. The traditional agricultural system of crop production in the most developing countries and Asia are based on grow- ing crops in mixtures, a system most commonly referred to as mixed cropping or intercropping.

Intercropping has long been recognized as a common practice among subsist- ence farmers in the traditional semi-intensive system of agriculture due to the flexibility of labour use and less risk. The concept brings together, crops with different maturity time, architecture, nutrient requirement etc into a mixture at the same time, under the same management for maximum utilization of nature for the benefit of man. When two or more crops are growing together, each must have adequate space to maximize complementarily between them. The predominance of intercropping, among tropical farmers cannot be overemphasized. The failure of different agricultural policies introduced by various governments in most of the developing countries to improve productivities was partly because most of the introduced modern technology adopted sole cropping. There has been a great awareness that most of these introduced technologies are contrary to the farmers' production practice.

As indicated the reasons why the peasant farmers in Africa are used to inter cropping are the several benefits which they derived. Among the economic and social benefits of intercropping are provision of favourable microclimate, possibility of lower labour requirement, stability of production, diversity of crops produced and higher productivity per unit area. Fortin and Pierce indicated that the interest in intercropping is growing worldwide because of its potential for increasing whole field productivity.

Good Agricultural Practices are collection of principles to apply for on-farm production and post- production processes, resulting in safe and healthy food and non-food agricultural products, while taking into account economical, social and environmental sustainability. Good agricultural practice may be applied to a wide range of farming systems and at different scales. They are applied through sustainable agricultural methods, such as integrated pest management, integrated fertilizer management and conservation agriculture.

5.3 CULTIVATION OF MEDICINAL PLANTS

Collection is an agricultural system in which uncultivated plants are harvested regularly or irregularly in the wild. Collection, simple processing and trading of medicinal plants contribute significantly to the cash income of the poor small holder farmers and especially women in developing countries. Studies carried out in India, indicated that from a single district of Pithoragarh in Uttranchal State of India, more than 1300 tons of medicinal and aromatic plants (MAPs) are collected and traded annually, most of them illegally. Unsustainable and large scale harvesting of MAPs from the natural habitats without providing equitable benefit to the local people and government is of grave concern to all. Cultivation of medicinal plants has pharmacological advantages over wild-collection. Wild collections normally vary in quality and composition due to environmental and genetic differences. Cultivation also greatly reduces the possibilities of misidentification and adulteration especially endangered species.

In many developing countries, medicinal plants were collected from the wild vegetation. However, in response to the combined impacts of dwindling supplies due to overexploitation of the natural resources and increasing demands due to population growth and growing global markets, threat from increase in human population leading to increase in human activities such as commercial agriculture, urbanization, human disturbance and deforestation, many plants species are facing the threat of extinction. Wild habitat became locked in parks and reserves making it difficult for people to collect the plant materials which they need until they search for a long time within long distance. In Ghana the activities of plant collectors have revealed that plant species that used to be obtained within short walking distance now have to be sourced several kilometers away owing to development of human settlements, farming activities, bush burning and other destructive human activities.

As seen, about 2,700 collections of medicinal plants in India, recorded 2,500 species of plants in Africa which have been found to have medicinal application in addition to another use. All these collections, the largest number are gathered from the wild while only a few are cultivated to a noteworthy extend. The intensive harvesting of wild medicinal plants due to the increasing use has in many places resulted in overexploitation, and forms a serious threat to biodiversity in the region. Several studies attest to a trend of increasing harvesting pressures on traditional supply areas linked to a growing shortage in supply of popular medicinal plant species. The current demand for certain taxa exceeds supply, with traders reporting acute short ages and price increases of these. As a result, several plant species have been exploited to such an extent that they are seldom found in un- protected areas.

The best way to provide the plants needed for medicinal purposes is to cultivate them. This is far better than collecting them from the wild since it thus not deplete the wild stocks, the declining habit of native plants can no longer sup- port the expanding market for medicinal plant products. In the case of rare, endangered or over-exploited species, cultivation is the only way to provide material without further endangering the survival of those species.

5.3.1 Bringing medicinal plants into cultivation

As observed, in America, the use of plants as medicine was formerly considered a primitive concept unacceptable in a modern nation with an enviable medical system, a reconsideration of the contributions that medicinal plant material can offer a contemporary health system has led to fresh views about medicine and plants. This reassessment of medicinal plants has stimulated cultivation of new species and led to increases in production of previously cultivated traditional species. As reported, more than 80% of the world's population in developing countries depends primarily on herbal medicine for basic healthcare needs; the use of herbal medicines in developed countries is growing and 25% of the UK population takes herbal medicines regularly. Around two-thirds of the 50 000 different medicinal plant species in use are collected from the wild, and in Europe only 10% of the medicinal species used commercially are cultivated.

As observed, traditional medicine is an important part of healthcare in southern Africa. Most drugs are prepared from plants, all of which are collected from the wild. All these activities on medicinal plants may lead to diminishing populations, extinctions and habitat degradation as many as 10000 medicinal species may be endangered in Europe.

It is imperative to develop the tradition of cultivating medicinal plants within the traditional cropping system to avoid extinction of most of the plants. There should be a holistic approach to develop agronomic practices within the traditional cropping system also to coordinate the local communities in establishing nurseries where the medicinal plants can be raised for commercial purposes. Although research in medicinal plants in Africa has generated a lot of information, information on the in- corporation of medicinal plants into the traditional cropping system and the evaluation of the therapeutic values of these plants under cultivation are scattered and their impact on health and economic remain limited.

Cultivation of medicinal plants especially high value medicinal plants is creating new dimension in the field of agriculture. To improve and conserve medicinal biodiversity, India has been reported to be making efforts for cultivation, collection, characterization, planned hybridzation programs for improvement of the plant species and conservation medicinal plants by restoring their genetic resources for commercial cultivation. Plant breeders of medicinal plant play key roles in improvement of medicinal plants because they care for stable produce, uniform plant growth and maturity, economic to grow and of good quality.

5.3.2 Inclusion of Medicinal Plants in cropping Systems

Some of the practical applications integrating medicinal plants into traditional farming systems have taken an obligate relationship in backstopping upland agriculture. South Asian states have a tradition of practicing mixed farming systems that include herbal plants and therefore, cultivation of medicinal plants especially applying organic and certified farming concepts has a greater scope in the region. Other important opportunities and advantages of cultivating MAPs include ease of their incorporation in the existing cropping systems due to availability of a large number of species and choice of plant types such as trees, shrubs, forbs, vines and their suitability to grow in different eco-physical conditions. Cultivation of carefully selected species as a mixed, inter or companion crop in agro and farm forestry conditions following a soil-improving crop rotation is highly feasible livelihood enhancing activities in South Asia. However, this will require an improved input and service delivery system including marketing, and post harvest technologies. Cultivation needs to be done on a business platform by a chain of small and microenterprise-based groups and individuals. In order to achieve an economy of scale and desired impact, it may need to be concentrated in selected pockets in an intensive manner as cluster of activities and micro- enterprises.

In cultivation, the variation and the resulting of uncertainty therauptic benefit is much reduced. There is no reason why traditional crop- breeding methods should or could not be applied to medicinal herbs grown domestically. The basic principles which include selection, multiplication and breeding only from individual genotypes showing the desired characteristics could be further hybridized to make crosses between specific genotypes, bearing in mind the desired combination of characters and an understanding of genetics.

Direct manipulation of DNA sequences to alter gene expression in medicinal plants is an area ripe for expansion. The primary target for trait manipulation in medicinal plants is the content of active compounds, but to develop them as crops basic agronomic characteristics related to uniformity, stability, growth and development, and resistance to biotic and abiotic stresses also need to be improved.

The commercial viability of bringing medicinal plants into domestic cultivation and the potential for increased use of modern biotechnologies are likely to be strongly influenced by popular perceptions both of herbs and of biotechnology. One of the main attractions of herbs as medicines is their 'natural' status and the associated, but erroneous, view that they must therefore be safe and intrinsically good for us. In stark contrast is the popular view of crops bred with the assistance of molecular biology and modern farming methods as highly 'unnatural'. This is particularly so for transgenic plants, and it is probable that organic growing methods will be received favourably by purchasers of cultivated medicinal plant extracts. However, it is entirely possible to grow selected or genetically modified varieties using organic growing methods.

5.4 CHALLENGES IN CULTIVATION OF MEDICINAL PLANTS

As observed danger involved if medicinal herbs are increasingly brought into domestic cultivation, then wild harvested plants would enjoy an increased cachet and commercial value, and non-sustainable harvesting methods would continue. It is perhaps worth considering that any form of cultivation or wild harvesting of plants is bound to involve the application of selective forces, conscious or unconscious, on the part of the grower or harvester. In the case of wild harvesting, continual selection of the largest wild-growing individuals or those with the traits considered desirable from a medicinal point of view, if it involves destruction of the whole plant or its reproductive organs, will inevitably lead to a degradation of the wild population. Though, it is difficult to use medicinal plants on a large scale without modifying the characteristics of the plant populations available, be they wild or domesticated. If medicinal herbs are brought into cultivation, then it will at least attempt to modify the characteristic of the population in a controlled fashion, and at the same time attempt to conserve wild populations.

Medicinal plants have always been part of mans' life on earth and there is a close relationship between plants and human beings. Man depends on plants for his existence and this relationship must be sustainable. Herbs have saved man's life on innumerable occasions and for these reasons their over-use concerns everyone. Plant-derived medicines have greatly provided the models for, a significant proportion of western drugs. Commercially-proven drugs used in modern medicine were initially used in crude form for other purposes that suggested potentially useful biological activity for centuries; the forests have been the source for procurement of medicinal plants. During the last few decades the area under forests has considerably decreased while the demand for raw material of medicinal plant origin has been increasing due to increased number of users due to resurgence of public interest in Homoeopathy. Therefore, the forest resources are not able to meet the full requirement of medicinal plants and herbs. Further, as a result of continuous exploitation of these plants in forests and absence of major regular development programs in the forestry and agriculture sectors, a number of species of these plants are reported vulnerable to extinction, and various sorts of prohibitions have been placed for use of a number of medicinal plants even for medicinal use.

5.5 PRODUCTION TECHNOLOGY OF ASHWAGANDHA, ASPARAGUS, ALOE, COSTUS, CINNAMOMUM, PERIWINKLE, ISABGOL

5.5.1 Ashwagandha

Introduction

Ashwagandha (Withania somnifera), also known as Indian ginseng, Winter cherry, Ajagandha, Kanaje Hindi and Samm Al Ferakh, is a plant in Solanaceae or nightshade family. It grows as a stout shrub that reaches a height of 170cm. Like the tomato which belongs to the same family, ashwagandha bears yellow flowers and red fruit, though its fruit is berry-like in size and shape. Ashwagandha grows prolifically in India, Pakistan, and Sri Lanka.

Ashwagandha possesses anti-inflammatory, antitumor, antistress, antioxidant, immunomodulatory, hemopoetic, and rejuvenating properties. It also has a positive influence on the endocrine, cardiopulmonary, and central nervous systems. The Indian Central Council for Research on Ayurveda and Siddha tells us that ashwaganda root exerts its most powerful pharmacological influence on the reproductive system, neuropharmacological disorders, cardiovascular system, respiratory system, bacteria, fungi, inflammation and gastric acidity

Ethanolic or aqueous extract of the root alone or in combination with other herbal materials is used in many commercial herbal formulations under various trade names such as Mentat, Geriforte, Pepcaps, 30 plus, Stresscom, Ashwagandharisht, Ashwagandhadi-churan, Ashwagandha-rasayanrist, and Ashwagandha-ghrit.

Ashwagandha has been used as an herbal remedy for hundreds of years. Ashwagandha has many applications, and has been used by Native Americans and Africans to treat inflammation, fevers, and to protect against infection or illness. Ashwagandha has also been used to boost the immune system, improve memory, and to promote overall wellness. Ashwagandha, winter cherry, is a shrub that flourishes in India and North America. The roots of the ashwagandha plant have been employed for millennia by Ayurvedic healers. Ashwagandha has many beneficial elements, including flavonoids and members of the withanolide class. Numerous modern studies have found that Ashwagandha shows great promise for being effective in reducing inflammation, decreasing stress, increasing mental activity, invigorating the body, and as an antioxidant.

Roots are used for medicinal purposes. Ashwagandha derives its name because of two reasons. First reason is that its root smells like horse and the other reason is that Ashwagandha vitalizes body to provide strength just like a horse possess.

Cultivation Process

Soil

Ashwagandha grows well in sandy loamy or light red soil, having pH of 7.5 to 8.0 with good drainage. Black soils or such heavy soils are suitable for cultivation.

Climate

Ashwagandha is grown as late rainy season (kharif) crop. The semi-tropical areas receiving 500 to 750 mm rainfall are suitable for its cultivation as rainfed crop. If one or two winter rains are received, the root development improves. The crop requires relatively dry season during its growing period. Ashwagandha can tolerate a temperature range of 20oC to 38oC and even low temperature as low as 10oC. The plant grows from sea level to an altitude of 1,500 meter above sea level.

Varieties

The Jawaharlal Nehru Krishi Vishwavidyalay has released one variety with high alkaloid, "Jawahar" which is short in stature and most amenable for high density planting. The variety yields in 180 days and yields a total with anolides content of 0.30 per cent in dry roots.

Plantation

Ashwagandha is a shrub, normally found everywhere in India especially in northwestern part of the sub continent, but is now a days cultivated in fields because of its great commercial demand. Plant size varies from 1 to 5 ft.

Nursery raising

Ashwagandha is propagated by seeds. Fresh seeds are sown in well prepared nursery beds. Although Ashwagandha can be sown by broadcast method in the main field, transplanting method is preferred for better quality and for exports. For export, a well maintained nursery is a prerequisite. The nursery bed usually raised from ground level is prepared by thorough mixing with compost and sand. About 5 kg of seeds are required for planting in 1 hectare of the mainfield. Nursery is raised in India, in the month of June or July. Seeds are treated in carbendazim to control wilt and seed borne diseases. Seeds are sown just before the onset of monsoon and covered thinly using sand. The seeds germinate in 5 to 7 days. About 35 days old seedlings are transplanted in the main field.

Field Preparation

2 to 3 ploughing and discing and / or harrowing should be done before rains. The land is well ploughed and pulverized and brought to a fine tilth. 10 to 20 tons of farm yard manure is applied. Field is then levelled.

Transplanting

After the manures are incorporated in the soil, ridges are prepared at 60 cm spacing. Healthy seedlings are planted at spacing of 30 cm. In some places, 60 cm x 60 cm or 45 cm x 30 cm spacing is also followed. However, a spacing of 60 cm x 30 cm with a plant population of about 55,000 seedlings per hectare is considered optimum.

Seed Requirement and Sowing Method

A seed rate of 10 to 12 kg per hectare is sufficient for broadcasting method. The seeds can be sown in lines also. Line to line method is preferred as this method increases root production and helps in performing intercultural operations smoothly. The seeds are usually sown about 1 to 3 cm deep. Seeds should be covered with light soil in both the methods. Line to line distance of 20 to 25 cm and plant to plant distance of 8 to 10 cm should be maintained. According to soil fertility, in fertile soil, distance can be extended.

Treatment of Seed

Seed should be treated with thirum or dithane M45 (Inofil M45) at the rate of 3 gms / kg of seeds before sowing to protect the seedlings from the seed borne diseases.

Thinning and Weeding

Grown up seedlings raised by sowing by broadcasting method or in line in furrows, should be thinned out by hand 25 to 30 days after sowing the seeds, to maintain a plant population of about 30 to 60 plants per square meter. The plant density to be maintained finally may depend on the nature and fertility of the soil. On marginal soil, the high population is maintained. If some fertilizer is applied, the population should preferably be kept at lower level. Generally two weedings are required to keep the field free from weeds, the first within 20 to 25 days of sowing the and the other after 20 to 25 days of the first weeding.

Manures and Fertilizers

The crop does not require heavy doses of manures and fertilizers. Ashwagandha responds well to organic manures and addition of 10 tons FYM and 1 ton of vermicompost per hectare is recommended. Application of 15 kg of Nitrogen and 15 kg of Phosphorous per hectare is beneficial for higher production.

Irrigation

Excessive rainfall or water is harmful for this crop. Light shower after transplantation ensures better establishment of seedlings. Life saving irrigation may be applied, if required. Under irrigated conditions, the crop can be irrigated once in 10 days.

Pests and Diseases

No serious pest is reported incase of this crop. Whenever the crop is damaged by insects, 2 to 3 sprays of rogor or nuvan should be applied @0.6%. A combination of 0.5% malathion and 0.1% to 0.3% kelthane as foliar spray at 10 to 15 days interval, was found highly useful for aphids, mites and insect attack. Diseases like seedling rock and blight are observed. Seedling mortality becomes severe under high temperature and humid conditions. Diseases can be minimised by use of disease free seeds and proper seed treatment, before sowing as stated earlier. Carbofuran should be applied @2 to 2.5 kg / hectare at the time of sowing. Neem cake also can be applied. Neem Cake will save root damage done by nematodes and insects. Further, adoption of crop rotation, timely sowing and keeping field well drained also protects the crop.

Harvesting

Maturity of the crop is judged by drying out of leaves and yellow-red berries. Flowering and bearing of fruits starts from December onwards. The crop is harvested for roots by digging in January to March, 150 to 180 days after sowing. There should be moisture in soil at the time of digging. Roots are dug out or ploughed using power tiller or a country plough. The tap root should be carefully pulled out, without damaging even the small lateral roots.

5.5.2 Asparagus

Introduction

Asparagus is a perennial crop that lends itself well to small-scale and part-time farming operations. Multiple markets exist for growers with 5 acres or less, and many field operations that require machinery, such as land preparation and planting, can be custom hired.

Asparagus is a member of the lily family (Liliaceae) and is one of a few vegetables that are monocots (plants having only one cotyledon, or seed leaf). Both male and female flowers are produced on the older asparagus varieties, but there are very few to no female flowers produced on the newer all-male hybrid varieties.

Generally, spears are harvested when they are 7 or 9 inches in length and green in color. Varieties with purple spears have also been developed by plant breeders. Excluding light when spears are emerging will produce blanched, or white, spears.

Asparagus is believed to be indigenous to parts of Russia, the Mediterranean region, and the British Isles. It was first cultivated by the early Romans, who used the asparagus for food and medicinal purposes. It was cultivated in England over two thousand years ago and brought to America by the early colonists. However, asparagus was not extensively planted by commercial growers until after 1850.

Most of the asparagus harvested in the United States is sold as fresh produce. The United States produces around 25,000-30,000 acres of asparagus with a value of $80-100 million. U.S. acreage is currently only about one-third of what it was 15 years ago due to increased imports from Central and South America.

Cultivation

Site Selection

Asparagus should be grown on well-drained soils that have good water-infiltration rates and good moisture-holding capacity. The soil should not be compacted and the pH should be 6.2 to 7.0. Growers should avoid planting asparagus in fields where it has been grown in previous years. Asparagus is an allelopathic species--it produces and releases toxic chemicals that inhibit and suppress the growth of young asparagus transplants or crowns. In addition, asparagus is extremely susceptible to Fusarium root rot, a soil fungus that will weaken the plant. Fusarium can survive up to seven years in infected soil and soil fumigation is not effective in reducing long-term Fusarium populations in the soil. Asparagus is extremely salt tolerant.

Climate and soil

The plant prefers annual average rainfall of 600– 1000 mm or less, of which 85% is received during July to September. A well-drained fertile sandy-loam to clay-loam soil, with a pH of 6–8 is best suited for its cultivation with staking support. Shatavar can be grown in open land as well as under shade, but very high moisture levels result in rotting of root.

Field Plantation

Land preparation and fertilizer application

The land should be given a deep disc ploughing, followed by harrowing and levelling. The field is normally divided into plots, keeping one irrigation channel in between two rows of plots. Ridges and furrows are made about 45 cm apart in the plots. About 10 tonnes of well-decomposed FYM is thoroughly mixed in the soil one month before transplanting. Shatavar further requires a fertilizer dose of 60 kg nitrogen, 40 kg phosphate, and 40 kg potash per hectare for optimum growth and higher tuberous root yield. One-third of nitrogen and entire dose of phosphate and potash should be placed 10–12 cm deep in the rows before transplanting.

Transplanting and optimum spacing

The seedlings are ready for transplanting after 45 days of sowing. These are transplanted in field at the onset of monsoon in July. Ridges and furrows are made 45 cm apart and seedlings are transplanted on ridges, keeping the plant-to-plant distance at 15 cm. The ridge method of transplanting is superior in comparison to flat method. Optimum number of seedlings required per hectare using the recommended spacing is about 150 000.

Intercropping system

Shatavar is normally grown as a monocrop, but it can be grown in inter spaces available in orchards having low light interception. Plants need staking material, thus poles or shrubs serve for support.

Interculture and maintenance practices

The balance two-third N is applied in two equal split doses at ridges during September and in late February. The fertilizer is broadcast in between the rows and mixed in soil followed by irrigation, if the soil is dry. Shatavar initially grows slowly for 60 days, which keeps inter-row space virtually vacant, allowing easy weed growth. It is necessary to carry out three weeding and hoeing operations to keep the field free from weeds for initial twomonth period. After two months, Shatavar grows enough to cover the inter-row spaces and prevents weed growth.

Irrigation practices

Water requirement of Shatavar crop is not much. It can be grown without irrigation in areas that receive 800–1200 mm of well-distributed rainfall. Irrigating the field once immediately after transplanting is a must for establishment of seedlings in field. The second irrigation is done after seven days of seedling establishment. If there is no rainfall and dry spell prevails for more than 15 days, one more irrigation should be given. During winters, irrigation at 30-day intervals is enough for good growth. Irrigation should be done during seed formation stage and before harvesting of the tuberous roots for obtaining higher seed yield and easy digging of tuberous roots. Deficient soil moisture during March–June brings down root yield significantly. Hence, three to four irrigations during this period are essential.

Pest Control

Weed control can be achieved with a good crop rotation system, herbicides, and straw mulch. Several preplant and postemergence herbicides are available for asparagus, depending on the specific weed problem and the time of year. If infestation levels are light, early cultivation (prior to spear emergence) can help reduce weed problems. Insects can be a major problem in asparagus production. Asparagus beetles, asparagus aphids, cutworms, and Japanese beetles all can cause crop losses. Monitoring insect populations will help you determine when you should use pesticides and how often you should spray.

Environmental Impacts

In the normal course of operations, farmers handle pesticides and other chemicals, may have manure to collect and spread, and use equipment to prepare fields and harvest crops. Any of these routine on-farm activities can be a potential source of surface water or groundwater pollution. Because of this possibility, you must understand the regulations to follow concerning the proper handling and application of chemicals and the disposal and transport of waste. Depending on the watershed where your farm is located, there may be additional environmental regulations regarding erosion control, pesticide leaching, and nutrient runoff. Contact your soil and water conservation district, extension office, zoning board, state departments of agriculture and environmental protection, and your local governing authorities to determine what regulations may pertain to your operation.

Commercially, asparagus can be started in the greenhouse 8 to 10 weeks prior to transplanting in the field or planted as one- or two-year-old crowns. Crowns are developed root systems with a fairly defined storage organ and growth buds.

Growers generally plant approximately 12,000 to 14,000 plants per acre in single rows, with 12 inches between plants in the row and 5 to 6 feet between rows. Whether planting crowns or transplants, the asparagus is planted in an 8-inch-deep furrow with a W-shaped configuration at the bottom of the furrow. The crown and transplant are planted in the W-shaped furrow beneath the soil surface, and the furrow is gradually filled with soil during the growing season. Asparagus usually is planted in May so that extensive foliage (fern) develops before winter.

Fertilizer recommendations should be based on annual soil test results. In absence of soil test results, the recommended N-P-K application rates are 50-100-150 pounds per acre broadcast in the spring of every year before spear emergence.

Harvest management

Crop maturity and harvesting

The crop matures in 12 months after planting; however, for seed harvesting, it is recommended to be harvested only after 20 months. Rabi season, that is, November–December, is the best time for harvesting tuberous roots when the above-ground parts start turning pale yellow. The crop, when harvested in 12 months, yields about 4–5 tonnes/hectare, while harvesting after 20 months yields about 6 tonnes/hectare of tubers along with 35 kg/hectare of seeds, which are not obtained in the 12-month-old crop.

Post-harvest management

After harvesting, the tubers are washed well in running water thereafter, these are dried in open sun for one to two days. The tuberous roots are then kept in luke warm water for one hour to soften the outer covering of the tubers. It facilitates removal of outer skin. The harvested roots are peeled manually by pulling their outer thin covering. These peeled tubers are then kept in shade for four to five hours, followed by further drying at 40 °C in hot air oven for 20 minutes or more, depending upon their moisture content. The roots should be completely dry for storage. If the tuber breaks with a cracking sound, it means that it has completely dried. Dried tuberous roots are packed in cardboard boxes and stored. Boiled tubers turn yellowish and fetch much higher market rate. Dry root in pieces of 5–15 cm × 1–2 cm are marketed as 'A' grade variety.

Chemical constituents

Sapogenins are the active principles found in the tuberous roots. Chemical evaluation studies suggest that the sapogenin content varies in the range of 0.7%–0.9% in the tubers.

Yield and cost of cultivation

The tuberous root yields 10%–12% of dry matter after removal of outer wall. An average shade-dried tuberous root yield of 3 tonnes per hectare is obtained from 20-month-old plants underexperimental conditions. A two-and-a-half-year-old crop gives a dry yield of 4–5 tonnes per hectare. It gives a seed yield of 30–35 kg/hectare. The estimated cost of cultivation is about Rs 100,000 per hectare, including land preparation, nursery raising, cost of planting material, FYM, cost of transplanting, harvesting, peeling and processing for market, and so on. The calculated yield of 20 month-old crop of Shatavar is 3 tonnes/hectare, which can fetch net returns of approximately Rs 62,500 /hectare. Dry root in 5–15 cm × 1–2 cm pieces are marketed as 'A' grade variety.

5.5.3 Aloe

Introduction

The name Aloe is derived from the Greek word for the dried juice of aloe leaves, and ferox refers to the spiny edged leaves. Aloe ferox is a succulent plant belonging to the Aloaceae family, including the dwarf aloes (Haworthia, Poelnitzia and Astroloba) and gasterias (Gasteria), which are also aloe-like in appearance and growth. The genus Aloe Linneus, with around 360 species, occurs predominantly in Africa with centres of species richness in South Africa, East Africa and Madagascar. Many of the species are woody and branching. The Aloe derives its name from the ferocious thorns (ferox in Latin) that cover the leathery surface of the leaves.

Aloevera (Aloe barbadensis) is a popular medicinal plant. It belongs to liliaceae family. It is a perennial plant, growing to the hight of 1½ - 2½ ft. Its leaves are long and thick, juicy with a wheel like phylotaxy. The two sides of the leaves have thorny structure with a thorny tip. The inner substance of the leaves is jelly like, with bad odour and bitter in taste. The length of the leaves ranges from 25-30 cm., while the breadth ranges from 3-5 cm. Normally it flowers during October to January and the long inflorescence has a large number of small pink flowers all around. Fruits are developed during February to April. It is normally not propagated through seeds. Vegetative propagation is easy and convenient. Of late, because of sky rocketing price of allopathic medicines with its known side effects, medicinal plants and ayurvedic medicines are becoming popular. World trade worth about 80 million US$ dollars exists now and this is likely to increase by 35-40 percent within 5 years. USA dominates the market (65%) while India and China have a share of 10 percent each which could be enhanced by its commercial cultivation.

Propagation

There are different methods that are used when propagating Aloe.

Seed

The plant germinates well from seed. Propagation of seed is relatively easy and should be used more often, particularly to ensure the survival of some of the rare species. In addition, it is very rewarding to grow it by this method.

Offsets or cuttings

Propagation by cuttings is the most popular method. The side branches or basal sprout are removed and allow the wound to dry off for a time.

Soil preparation

The field should be prepared well before heavy rainfall to achieve a fine tilth.

Field layout and design

A bed width of 0,9 m and a furrow of 0,3 m summing up to 1,2 m is recommended. The bed height should be maintained at about 15 to 20 cm. Accommodate two rows on each bed, maintaining a distance of 60 cm between rows and 50 cm between plants. This planting pattern will give an average plant population of 33 333 plants/ha with an effective distance between two laterals of 1,2 m. The drip line should be placed between two rows on each bed. Adopt an emitter spacing of 0,5 cm.

Planting

Aloe can be planted in spring, 1, 5 to 2 m from each other in the nursery. When the seedlings have three or four leaves, or are about 3 cm tall, they should be planted into 1-kg bags containing a well-drained mixture of sand and compost and after two years they can be planted out into open ground. About 15 to 18-cm long root suckers or rhizome cuttings should be planted in such a way that a two-third portion of the planting material is underneath the soil.

Fertilisation

A small quantity of manure is required to enhance the growth of the plant. Organic compost can also be used when making planting holes to speed up growth.

Irrigation

Although the plant is drought tolerant, it thrives and flowers better if adequate irrigation is provided in the summer months. The land should be irrigated immediately after planting.

Weed control

The land should be kept weed free by weeding the fi eld when necessary.

Pest control

The major insects identified in Aloe include aloe snout beetle, scale insects, mealy bug and mites.

Cultivation Factors

Soil and Climate

Aloe vera is found to grow in hot humid and high rainfall conditions. It is grown in all kind of soils but well drained soil with high organic matter, is most suitable. It grows well in bright sun light. Shady conditions results in disease infestation It is highly sensitive to water stagnation. Therefore, well drained high land should be selected for its cultivation. A rainfall ranging from 1000 - 1200 mm is ideal for aloevera cultivation. Seedling Preparation and Planting Since it is difficult to grow aloevera from seeds, seedlings are normally raised from roots of the plants. Sucker itself can be used as seedlings as in Banana. Rainy season is ideal for sucker plantation. A spacing of 1.5 x 1 ft, 1 ft x 2 ft or 2 ft x 2 ft is followed. Land Preparation About 2-3 ploughings and laddering are done to make the soil weed free and friable. Land leveling is then followed. Along the slope, 15-20 ft apart drainage are made.

Fertility of soil

Since aloe vera can be grown in any type of land, it is the fertility factor which determines the growth of the aloe vera plant. Tests your land fertility through govt. Lab, testing of soil will make you sure of about how much your soil is fertile and according to that, you can give doses of N: K: P, and increase the growth of aloe vera plant.

However, 3 to 4 tons of farmyard manure (FMY) is necessary for the better growth of the aloe vera plant along with 25 kg urea, 35 kg Phosphorus and 10 kg of Pour potash per acre.

Give nitrogen three times and phosphorus and potash at the time of preparation for the land. It is good to sprinkle nitrogen on plants for the health of aloe vera plant at the time of its growth.

Irrigation

Aloe vera plant needs very less water to grow but at a regular period as per their requirement which will depend on the climate condition. Usually, first water is given at the time just after its planting in the farm which helps the newly planted aloe vera to set well.

Then after its requirement, water is to be done at every regular interval. At each year, three to four-time watering is to be done so that aloe vera plant can develop well during the complete year.

Drip irrigation and sprinkler irrigation are the best way for the farming of aloe vera because it saves water along with the precious time. This will also make you more profitable by increasing the yield of aloe vera. Research on aloe vera was stated that time to time irrigation increase the production, quantity, and quality of the aloe vera and give a better price in the change of your yield.

Drip irrigation and sprinkler irrigation also reduce the laborer costing needed for the watering in the farming. Since aloe vera plants are very sensitive to the water, over irrigation then the requirement of plant leads you to defect in the farming as result there is a decrease in the production of the aloe vera farming.

6.5.4 Costus

Introduction

In India, the agroclimatic conditions provide an ideal habitat for the growth of more than 9500 medicinal plants. Thus, there is an enormous scope for India to emerge as major player in medicinal plant sector. Plant products contain many important chemicals with medicinal and valuable pharmacological properties. There is a growing demand for natural products of medicinal/pharmaceutical importance in both domestic and international market. The current demand for the plant based products in medicine and industry has resulted in extensive investigation of the plants for potential therapeutic agents. CostusLinn. is a tropical herbaceous plant from family Costaceae under the order Zingiberales. The leaves are large and spirally arranged on stems which differentiate it from its nearest relative, the zingiber or true ginger. Hence these plants are often referred as the spiral ginger or Crepe ginger. The Costus spp are commonly grown as medicinal and ornamental plants. The rhizome is the major source of diosgenin, which is anti-diabetic in nature and is used in the treatment of diabetes mellitus. The most popular species in the genus is C. speciosus, which has emerged as an important antidiabetic plant

The different species of Costus vary in flower colour. Some varieties with flowers and bracts look like compact cones, while others are shaped like pineapple or soft crepe coming out of green cones. Some leaves are pubescent on abaxial surface, while others are smooth and purplish. About seven species of the genus Costus Linn. are known from India. Other cultivated species of this genus are C.barbatus, C.chartaceus, C. cuspidatus , C. giganteus, C. igneus ,C. osae, C. spectabilis.

Cultivation

C. speciosus are usually grown in fertile, organic, moist, well-drained soils in shade. Tropical climate with high humidity and minimum temperature 13o C is best for its cultivation. Crepe ginger grows from thick fleshy roots called “rhizomes”. A single rhizome will produce new shoots and increase to a 3 ft wide clump in the second year under ideal growing conditions. Costus reproduces vegetatively by rhizomes, division of culms, stem cuttings. It can also be grown through seeds. Costus seeds are dispersed by birds when they feed on the fruits. The percentage of seed germination is about 62%. Standard cultivation practices of C.speciosus have yet not been developed.

Propagation

- ☆ Although the plant can be propagated from seeds, stem cuttings and rhizomes
- ☆ Commercially it is being propagated only through rhizome cuttings.
- ☆ Selection of rhizomes for planting is however important. The rhizomes have a number of nipple shaped buds most of them being concentrated around the stem scars and the tips.
- ☆ The formation of buds on the rhizomes is poor during April. Cuttings of rhizome pieces for propagation should have at least two viable buds. The rhizome pieces weighing around 40g should be selected.
- ☆ It can also be grown through seeds. Costus seeds are dispersed by birds when they feed on the fruits. The percentage of seed germination is about 62%.

Seed rate:

2000 to 2400kg/ha. Rhizomes required.

Land preparation

Land is ploughed 2-3 times and soil is brought to a fine tilth. FYM at the rate of 15 tonnes is applied and mixed well with the soil. Furrows are opened at 50 cm apart.

Planting

- ☆ The rhizome pieces are placed at a depth of 8-10 cm taking care to place the eye buds facing upwards, horizontally in rows at 50 cm apart and covered with soil.
- ☆ The crop is irrigated immediately after planting. Thick sized pieces show a slow rate of sprouting till 40-45 days of planting.
- ☆ This is due to eye buds being dormant on these rhizome pieces which take a longer time to develop especially in case of the crop planted during April. After 70-75 days about 90-95 percent sprouting is obtained
- ☆ Desiccation of young sprouts has been observed during hot summer months, necessitating liberal water supply during this period.
- ☆ Best period of planting is from 3rd week of April to 3rd week of May. Rhizomes planted during June and July complete germination within 60 days but yields

obtained are low as the crop has a shorter growth period in the first year before dormancy sets in.

- ☆ The active vegetative growth period of crop is from July to Mid September and maximum tuberization takes place between September to November before dormancy sets in.
- ☆ The effect of planting on ridges versus flat beds indicated that there was no significant difference in yield due to the method of planting although a higher yield of fresh rhizomes was obtained in the crop planted on ridges.

Manures and Fertilizers

- ☆ It is a rhizomatus crop and to compensate the biomass production, heavy manuring is required.
- ☆ In a trial conducted at Vallanikkara, the tentative optimum dose for obtaining the maximum yield of diosgenin was observed to be 45 kg N, 30 kg P2O5 and 30 kg K2O along with 15 tonnes of FYM per hectare.
- ☆ The farm yard manure and half dose of phosphorus and potash are applied in 2 split doses at 20 and 60 days of planting, the remaining half dose of phosphorus and potash is given along with the second dose of nitrogen after 60th day of planting.

Irrigation

- ☆ The crop requires a liberal supply of water for successful growth. The crop planted during April and May requires to be irrigated atleast two to three times a month till the outbreak of monsoon.
- ☆ As September to November is the period of miximum tuberization atleast two irrigations should be given.
- ☆ During the dormancy period (December to March), if winter rains are scanty, there is need for irrigation.
- ☆ If moisture conditions are adequate it has been observed that the crop sprouts towards the end of March or early April.
- ☆ With a liberal supply of irrigation during the hot summer months excellent crop growth is observed during these months.
- ☆ Thus, for a crop to be harvested after a period of 17 to 18 months, 14 to 17 irrigations are required.

Weeding and Inter-culture

- ☆ One weeding during the sprouting period of the crop followed by two more keeps the crop fairly free of weeds.

- During the period of active vegetative growth (July to September) most of the weeds are suppressed. If the monsoon is erratic, there is need to remove weeds at least once.
- Weeding once or twice during the dormant season helps in better sprouting of the crop during the next season.
- During the second year of the crop, the inter-row space between plant is covered up by the canes and only two to three weeding are necessary before the crop is harvested in August or September

Pests and Diseases

There are no major pests which affect this crop. However, following are some diseases which are observed to affect the crop.

1. Rhizome rot:

- It is caused by Phytophthora solani and develops very fast from July to August, in the rhizomes kept in storage after the harvest.
- Initially symptoms start from the tip from the injure portion of the rhizomes. Subsequently, the rhizome becomes light brown and gives offensive odour.
- Properly harvested rhizomes without injuries can be stored for long time. No chemical control has been worked out.

2. Pythium rhizome rot

- It is caused by Pythium spinosum.
- In this disease, the infection in the plant starts from injured portion of the rhizomes and in later stages rhizomes turn into dirty brown in colour, the leaves become yellowish brown and finally dry off.
- Selection of rhiozmes from the healthy crop and dipping the rhizomes in fungicidal solutions like Benlate, Bavistin or Dithane-z-78 has been recommended to control the disease.

3. Leaf blight

- It is caused by Curvularia paradisii. It is very severe from July to September.
- The symptoms develop on the leaf lamina as a small, spherical to irregular spots, light brown in colour, and in advanced stages they become dark brown upward.
- Curling of young leaves towards upward is very common. In advanced stages all the leaves fall down leaving bare stem in the field.

- ☆ This pathogen can be effectively controlled by spraying 0.3 per cent Dithane M-45 at fortnightly intervals.

Harvesting

- ☆ It is observed that diosgenin content is maximum when the crop is in an active stage of vegetative growth, and when it is of 16-17 months old.
- ☆ Two operations viz., harvesting the aerial shoots and digging out the rhizomes are required to complete the harvesting in costus.
- ☆ After harvesting the aerial shoots, the most satisfactory and economic method to dig out the rhizomes is to run the tractor drawn cultivator crosswise twice or thrice in the field and simultaneously collect the uprooted rhizomes manually.
- ☆ A negligible amount of damage is caused to rhizomes during the operations.
- ☆ The harvested rhizomes are spread out for a few days and the earth sticking on the rhiozmes is removed or alternatively the rhizomes are washed under the jet of water from the tube well discharge pipe.
- ☆ The rhizomes have to be chopped and dried before diosgenin can be extracted.
- ☆ Hand chopping is time consuming and costly, so chopping by mechancial means has to be resorted to.
- ☆ Several methods have been tried but the ordinary wheat theresher with a little modification is found satisfactory for chopping the fresh rhizomes.
- ☆ Very fine chopped material dries quickly (2-3 days)

5.5.5 Cinnamomum

Introduction

Cinnamon is the dried bark of the perennial tree of C.zeylanicum of the Lauraceae family. True cinnamon is native to Sri Lanka. Cinnamon is originally grown wild in central hill country of Sri Lanka. The history of cinnamon dates back to about 2800 B.C where it can be found referenced as "kwai" in Chinese writings. Cinnamon is even mentioned in the Bible when Moses used it as an ingredient for his anointing oil in ancient Rome. It was burned in Roman funerals perhaps partly as a way forward to off the odor of dead bodies. Emperor Nero is said to have burned a years' worth of the dry supply of cinnamon at the funeral of his wife Poppaea Sabina. Ancient Egyptians used it in embalming mummies because of pleasant odors and its preservative qualities.

Cinnamon was a precious spice in the west during 14th -15th centuries and its' primary use was to preserve meat and to retard the growth of bacteria. The quest for cinnamon was a major factor which led to exploration of the world in 15th century. By that time the real cinnamon was produced in only one place, namely in Ceylon or Sri Lanka. Anyone who had the control of the supply flow would have made profits immensely. Portuguese traders made their way to Ceylon in the 15th century, enslaved the natives and had the control of the trade from Arabs. Soon the Dutch displaced the Portuguese and gained the control of the cinnamon monopoly. It was the Dutch who took a massive effort to boost the production by domestication of crop and expanding extent in the areas they had the control. Because of that effort cinnamon cultivations were moved to Western and Southern coastal belts of the island. Since 1815 the British took the control of the island and cinnamon trade too was moved to their hands. By this time the relative importance of spices in the world market had been declining due to the emerging plantation crop sector of tea and rubber, which restricted the further expansion of cinnamon.

Varieties

There are eight cinnamon species in Sri Lanka. Among them only Cinnamomum zeylanicum Blume is grown commercially. In tradition, there were several types of cinnamon categorized based on taste of the bark. "Pani-Miris Kurundu" was the best with sweet-pungent taste and "Miris Kurundu", "Sevel Kurundu" and "Thiththa Kurundu" are the others. Currently ten cinnamon accessions have been identified based on yield and quality performances and best two lines, named as "Sri Vijaya" and "Sri Gamunu", were released. Other selections are under evaluation in different agro climatic zones.

Propagation

Two major methods of Cinnamon propagation

1. Seed propagation
2. Vegetative propagation

What ever the method used for the propagation, selection of best mother plant is the major factor under concern. Criteria for the selection are listed below.

- ☆ Erect stems with smooth bark
- ☆ Vigorous growth
- ☆ Easiness of peeling the stem bark
- ☆ Free from pest and diseases

- ☆ Good quality characteristics; sweetness, pungency, flavour etc.
- ☆ Dry bark yield (Current average yield in Sri Lanka 470/ha/yr)
- ☆ Oil yield (Leaf & Bark) (current average oil yield - 0.9% both leaf and bark oil where it could reach 3% in selected varieties)
- ☆ Oil Quality (Percentage Euginole in leaf oil and Cinnamaldihyde in bark oil)

1. Seed propagation

The collected fleshy berries are left in heaps for about 2-4 days in shade in order to soften and rot. Pulp is then removed and air dried. Over fermented and partially filled seeds along with damaged seeds should be excluded by hand sorting. Only the large and medium sized seeds are selected.

Category	Length	Width
Large	>12.5 mm	>9.9 mm
Medium	8.5 - 12.5 mm	6.1 - 9.9 mm
Small	<8.5 mm	<6.1 mm

Seed bed or Cinnamon ball method

In this method raised beds of about 1 m breadth and convenient length are prepared to facilitate good drainage. In order to provide the plants with good soil texture, soil is cultivated to a depth of about 30 - 40 cm. Planting holes of about 4 cm in diameter and 4 - 8 cm in depth are prepared at 10 x 20 cm spacing using a peg. After filling these holes with some fine structured soil and 7 - 10 seeds they are covered with a layer of soil and may be watered daily if necessary. Artificial shade should be provided until plants are about 12 cm tall and the removal of the shade also should be gradual. Seed germination starts in about 20 days and when the plants are about 3 - 4 months old weaker ones may be removed leaving 3 vigorous plants per point. In about 8 - 10 months plants are ready for planting on the field. It is important not to damage the roots when removing the plants from the nursery. Thoroughly watering before the extraction is the best solution for this problem and supporting the root ball with a thick mud slurry would be an added advantage if these plants are to be transported.

Raising seedlings in poly bags

Polythene bags of 12.5 cm x 20 cm size and thickness of 250 gauge with a potting mixture composed of equal parts of top soil and well decomposed cow dung also can be used to raise seedlings. At least 8 seeds per bag is sown and thinning out is done if necessary to keep 5 to 10 seedlings per bag). Field planting can be done after 3 months. potting bags can be kept for a maximum of 8 months if correct potting mixture is used.

2. Vegetative Propagation

1 Propagation by stem cuttings

Cinnamon can also be propagated by cuttings of young one-leaved shoots or by layering. Partially matured shoots (Semi hard wood) with a node are removed from selected mother plants with desired characteristics. Single node shoots are prepared by making sloping cut just above the node (length of cutting 1"). Cuttings should be put into water immediately and maintained in water until planted in polythene bags. Polythene bags 12.5 cm wide and 20 cm long should be pressed to be firm. The filled polythene bags should be put together, within frames made of bamboo or suitable supports to give beds not more than 1 m wide. The soil under the pots should have been forked over to ensure good drainage. 2-3 cuttings should be placed in each bag. The bed of polythene bags and cutting must be kept moist. In order to prevent water losses through evapo-transpiration the bags must be covered with polythene (This type of an arrangement is known as simple propagator). It is also important to protect from direct sun light. After 2 months, the shade has to be removed gradually for hardening of plantlets. The rooted plantlets are ready for planting in 6-8 months.

2 Propagation in vitro

In vitro propagation has the potential for rapid multiplication of selected plant species with desired characteristics. The possibility of using Plant Tissue Culture techniques for the rapid multiplication of Cinnamomum zeylanicum has been established. Ravishankar Rai and Chandra (1987) induced multiple shoots from hypocotyl segments of seedlings on Murashige and Skoog's basal medium supplemented with each ?-Naphthalene acetic acid (NAA) and 6-Benzylamino purine(BAP) at 0.5 mg/L. However there is no information about the adoption of Tissue Culture technology for commercial micro-propagation of Cinnamon so far, and is an area needing research and development.

3 Air Layering

The possibility of propagation of Cinnamomum aromaticum Nees (Cassia) through Air Layering has been recorded. According to Krishnamoorthy and Rema (1994) propagation of Cassia has been achieved through air layering with 50% to 87.5% success depending on the time of the year.

Factors

Soil and Climate

Soil

Cinnamon is a hardy plant which can grow on any soil under a wide variety of tropical conditions, but to derive the characteristic flavour, careful cropping conditions must be observed.

Cinnamon can be grown in various types of soils varying from silver sands in Negambo to loamy and lateritic gravelly soils in Southern costal belt and interior. The bark quality is influenced by soil and climatic factors and the best quality cinnamon is produced in white sandy soil in Negombo area. However, the best grade cinnamon called "Alba" is mainly produced in Red Yellow Podzolic soils in South-western region of Sri Lanka. Cinnamon needs a deep soil but cinnamon roots can penetrate even through the cracks of the parent material to deeper layers.

Climate

- ☆ Cinnamon is commercially grown in coastal belts in Sri Lanka and spread to interior part of the country where elevation is increased up to about 250 above MSL. Naturally cinnamon has been found in central hilly area of Sri Lanka the elevation increased up to about 500m amsl. Until to date it can be found in Sinharaja and Knuckles forest reserve. Wet zone is ideal for the successful growth of cinnamon but it can be grown commercially in Intermediate zones of mid and low country, where annual rainfall is more than 1750mm. However it is not suitable for areas with prolong dry periods.
- ☆ Cinnamon is sun loving plant and high sun shine is needed.
- ☆ The most suitable temperature is between 250C- 320C .
- ☆ Rainfall should be in the region of 1,750-3,500 mm per annum.

Field Planting

On flat land cinnamon can be planted in straight lines and on sloping land planting on the contour lines is recommended. Spacing can be adjusted according land type, 1.2 m x 1.2 m (Flat land), 1.2 m x 0.9 m (Flat land/Gentle slope), 1.2 m x 0.6 m (Steep lands). 0.3 m x 0.3 m x 0.3 m planting hole can be filled with top soil and preliminary research have shown that, addition of 25 g of Rock Phosphate seems to have a positive effect on root development (Further experiments are needed).

At 1.2 m x 0.9 m spacing one acre occupies 3600 planting points. Planting must be done at the on set of rains at a rate of 5 plants per hole. Planting holes could also be sown direct with prepared seeds during the rainy season. If the inter plant spacing is too close then results a situation where the plants tend to grow elongated seeking the sunlight. This impairs the eventual flavour characteristics of the bark.

Soil Management

When the land is sloppy adopting of soil conservation measures is very important factor.

Digging of contour drains at suitable distance depending on the slope and the rainfall. The drains should be deep enough to allow for settlement of eroded soil.

Mulching with pruned branches and weeds

Following recommended spacing

Burying of weeds and pruned leaves around the root stocks is beneficial. Gathering earth up to the root stock without mounting should be done as cinnamon is a surface feeder. The root stock should be exposed to sun light to allow new shoots to develop and to prevent termite attacking the root stock.

Fertilizers

Fertilizer application is important for commercial cultivation of cinnamon for higher productivity, as cinnamon is normally grown as a long-term monoculture. In the absence of fertilizer application, the supply of nutrients available in the soil will become exhausted leading to mineral deficiencies and drop of yield; dry matter yield as well as the oil content of the bark and leaves and poor stooling. The requirement of fertilizer may be based on the results of field experiments, soil and plant tissue analysis or the symptoms of mineral deficiencies or toxicities. As such the optimum applications will vary from one region, plantation or field to another in accordance with local conditions. The Department of Export Agriculture of Sri Lanka recommends following fertilizer mixture and quantities, for cinnamon in Sri Lanka.

Weeding

Weeding is also an essential operation in cinnamon. Clean weeding is recommended for young plantations and slash weeding is recommended at 2-3 times a year for mature crop.

Soil conservation

In areas where the land is sloppy or undulated, soil conservation reduce the erosion. Contour trenches at appropriate intervals are recommended.

Plant training and pruning

Training and pruning of plants should be done once in every six months. Excess lateral branches are removed to have a straight and smooth stem and after harvesting weak shoots are removed to enhance the growth of main stems.

Harvesting

Cinnamon is ready for harvesting after 2-3 years from planting when the plant reaches height of 1.5-2 m with three to four shoots and the bark turns brown in colour. The main shoot is coppiced or cut back to a height of about 6 cm from ground level. Two to three crops are taken annually depending upon the rainfall.

REFERENCES

Ahenkan, A. and Boon, P. 2008. Enhancing Food Security, Poverty Reduction and Sus- tainable Forest Management in Ghana through Non-timber Forest Products Farm- ing: Case Study of Sefwi Wiawso District, published online by GRIN Publishing at www.grin.com/de/preview/.html (Retrieved 15 May, 2010)

Alternative Medicine News 2008. India to Step up Allocation for Cultivation of Medicinal Plants Tuesday, May 13, 2008

Australian Centre for International Agricultural Research (ACIAR). 2007. "Research that works for developing countries and Aus- tralia". http://www.aciar.gov.au/publication/public ationsbydownload. Retrieved 25 November 2007.

Canter, P. H. 2005 Bringing medicinal plants into cultivation. Focus Altern Complement Ther 10: 167–168

Canter, P. H; Thomas, H. and Ernst, E. 2005. Bringing medicinal plants into cultivation: opportunities and challenges for biotechnology.

Craker, L.E. 2008. A perspective on cultivation of medicinal plants in America. Acta Horticulturae 765:57-66

Cunningham, A.B. 1991a. Development of a conservation policy on commercially ex- ploited medicinal plants: a case study from southern Africa. In: Akerele, O., Heywood,

V. and Synge, H. eds. The conservation of medicinal plants: proceedings of an inter-national consultation 21-27 March 1988, Chiang Mai, Thailand. Cambridge Univer- sity Press, Cambridge, 337-358.

Cunningham, A.B. 1991b. The herbal medicine trade: resource depletion and environmental management for a hidden economy. In: Preston-Whyte, E. and Rogerson, C. eds. South Africa's informaleconomy. Oxford University Press, Cape Town, 196–206.

Dixon, J., Gulliver, A. and Gibbon, D. 2001. Farming systems and poverty: Improving farmers livelihoods in a changing world. Food and Agriculture Organization, Rome, and World Bank, Washington, DC

Elemo, K.A. and B. B. Mabbayad 1980. Effect of sowing date on upland rice/peanut inter- action. The Philippine Journal of Crop Sci- ence 5 (1): 22-29

Fortin, M.C. and Pierce F.J. 1996. Leaf azimuth in strip-intercropped corn. Agronomy Jour- nal, Madison, 88(1): 6-9

Karki, M., Tiwari, B., Badoni, A. and Bhattarai, N. 2003. Creating Livelihoods Enhancing Medicinal and Aromatic Plants based Bio- diversity-Rich Production Systems: Prelim- inary Lessons from South Asia Oral Paper Presented at The 3rd World Congress on Medicinal and Aromatic Plants for Human Welfare (WOCMAP III) 3-7 February 2003, Chiang Mai, Thailand.

Kempanna, C. 1974. Prospects for Medicinal Plants in Indian Agriculture. World Crops. 26: 166-168.

Osiru, D.S.O. 1982. Intercropping: a review of the possible advantages. Proceedings of the Indian Statistical Institute International conference on Frontiersa of Research in Agriculture. Calcutta, September 27- October 1, 1982. pp 304-320

Pieters, L. and Vlietinck. A. J. 2005. Bioguided Isolation of Pharmacologically Active Plant Components, Still a Valuable Strategy for the Finding of New Lead Compounds? J. Ethnopharmacol. 100(1-2): 57-60.

Prasad, P. 2009. Impact of cultivation on active constituents of the medicinal plants Podophyllum hexandrum and Aconitum heterophyllum in Sikkim PGR Newsletter FAO Biodiversity 124: 33 – 35.

Rao, M. R., Palada, M. C. and Becker, B.N. 2004. Medicinal and aromatic plants in ag- roforestry systems.Agroforestry Systems, 61 (1), 107-122.

Regmi S. and Bista, S. 2002. Best Practices in Collection and Cultivation of Medicinal Plants for Sustainable Livelihoods in Hima- layan Communities; Paper presented at the Regional Workshop at Wise Practices and Experimental Learning in the Conservation and Management of Himalayan Medicinal Plant; December 15-20, 2002; Kathmandu. CECI- Nepal.

Schmelzer, G.H. and Gurib-Fakim, A. (Editors), 2008. Medicinal plants of Tropical Africa Conclusions and recommendations based on PROTA 11(1): 'Medicinal plants' CTA 791 pp

Vines, G. 2004. Herbal harvests with a future: towards sustainable sources for medicinal plants, Plant life Integration

INDEX

www.ingramcontent.com/pod-product-compliance
Ingram Content Group UK Ltd.
Pitfield, Milton Keynes, MK11 3LW, UK
UKHW021327190726
13845UKWH00004B/62

9 789349 964990